# THE COMPLEX WORLD OF ANTS

Edited by **Vonnie Shields**

**The Complex World of Ants**

http://dx.doi.org/10.5772/intechopen.70081
Edited by Vonnie Shields

**Contributors**

Raphael Vacchi Travaglini, Alex Vieira, Maria Izabel Camargo-Mathias, André Arnosti, Luiz Forti, Roberto Camargo, Luis Stefanelli, Elena Lopatina, Ganesh B Gathalkar, Avalokiteswar Sen, Vonnie D.C. Shields



**Notice**



First published in London, United Kingdom, 2018 by IntechOpen
IntechOpen is the global imprint of INTECHOPEN LIMITED, registered in England and Wales, registration number: 11086078, The Shard, 25th floor, 32 London Bridge Street
London, SE19SG – United Kingdom
Printed in Croatia

British Library Cataloguing-in-Publication Data
A catalogue record for this book is available from the British Library

Additional hard copies can be obtained from orders@intechopen.com

The Complex World of Ants, Edited by Vonnie Shields
p. cm.
Print ISBN 978-1-78984-267-8
Online ISBN 978-1-78984-268-5

For EU product safety concerns:
IN TECH d.o.o., Prolaz Marije Krucifikse Kozulić 3, 51000 Rijeka, Croatia, info@intechopen.com or visit our website at intechopen.com.

# Meet the editor

Vonnie Shields, PhD, is currently a full professor in the Biological Sciences Department and acting dean in the Fisher College of Science and Mathematics at Towson University, Towson, MD, U.S.A. Dr Shields' laboratory engages in multidisciplinary research directed toward exploring the importance of gustatory, olfactory, and visual cues in the selection of food sources by carrying out behavioral and electrophysiological studies on larval and adult insects. In addition, her laboratory examines the structural organization of insect sense organs using transmission electron and scanning electron microscopy. The overall goal of this research is to acquire a better understanding of the sensory mechanisms by which insects find host-plants and detect plant-associated volatiles. The aim is to discover possible novel biocontrol agents against insect pests. Dr Shields studied biology at the University of Regina, Regina, Saskatchewan, CA. Her interest in insect chemosensory research began after her undergraduate studies, when she started her PhD studies at the same institution. For her PhD, she carried out research at the University of Regina and the University of Alberta, Edmonton, Alberta, CA. After graduating, she accepted a research associate position to conduct postdoctoral studies at the Arizona Research Laboratories Division of Neurobiology, University of Arizona, Tucson, Arizona, U.S.A., before she accepted a faculty position at Towson University.

# Contents

Chapter 1

# Introductory Chapter: The Complex World of Ants

Vonnie D.C. Shields

Additional information is available at the end of the chapter

http://dx.doi.org/10.5772/intechopen.80387

## 1. Introduction

### 1.1. Origin, classification, and distribution

Ants fall into the largest family of insects, with respect to diversity of species and total number of individuals. They belong to a single family, the Formicidae, within the order Hymenoptera, consisting of 11 subfamilies and 297 genera [1]. In 1967, Wilson et al. obtained the first ant remains of the Cretaceous period [2, 3] that were estimated to be about 80 million years. Having said this, there are well over 8800 described species, to date. Most of the genera stem are from the Neotropics and Afrotropical regions. Remarkably, individual colonies may contain 20 million members [4]. Ants dominate much of the terrestrial regions of the world, ranging from deserts to subarctic tundra. Interestingly, recent measurements indicate that approximately one-third of the entire animal biomass of the Amazonian *terra firme* rain forest is composed of ants and termites. Each hectare of soil contains in excess of 8 million ants, in contrast to only 1 million termites [1].

### 1.2. Morphology

Ants (Formicidae) bear a narrow "waist" between the abdomen and thorax. The "waist" is comprised of a one-segmented (petiole) or two-segmented (petiole and postpetiole) constriction located between the thorax and bulbous poster portion of the abdomen or gaster. The gaster is composed of 4–5 posterior segments. Ants have large heads and powerful jaws. The adult workers and queens bear bent (elbowed or geniculate) antennae consisting of a long basal scape and 3–11 short funicular segments. In males, the basal segments are not long, so the antennae will not appear to be bent. The last two or three segments may be enlarged, forming a club. The antennae are believed to function as a two-way communication device [5], rather than just a receptor. Wang et al. analyzed the behavior and surface chemistry of many ants, focusing on the use and function of the cuticular hydrocarbons covering the ants' bodies.

They found that this layer not only served to protect the ants from dehydration, but also formed a critical role in their communication. Their research showed that when the layer was removed from the antennal surface, information regarding nest identification was lost, no longer allowing the ants to identify their colony.

### 1.3. Feeding behavior and communication

Ants play important roles in natural ecosystems. They are omnivorous feeders and can live in a wide range of habitats. They build nest sites, typically underground, thus contributing to nutrient cycling, seed dispersal, and the scavenging of dead organisms. They are considered major predators of other arthropods and small invertebrates, including pests, such as crop feeding caterpillars and ticks. Secretions stemming from their metapleural gland, rich in antibiotics, allow them to disinfect moist environments and allow them to live in areas that other organisms do not live in, especially in the tropics [1].

Communication is necessary in order to coordinate the activities within a colony. This is mediated via chemical signals (pheromones). Some of these pheromones include a queen pheromone, allowing workers to recognize a queen, trail following pheromones (used by workers to mark paths between the nest and food), and alarm pheromones (cause ants to disperse and/or attack). In addition, chemical cues are used to recognize colony nestmates and play additionally a role in aggression and territorial boundary markings between colonies [1].

### 1.4. Life/colony cycle

Ants are eusocial insects. They (1) perform cooperative brood care, where the adults care for the immatures, (2) bear overlapping generations, and (3) have a division of labor among reproductive and nonreproductive (workers) groups. The latter group is responsible for performing tasks necessary for colony survival, including foraging, care of immatures and reproductive adults, and nest building. Such a division of labor results in the formation of castes or specialized behavioral groups [1].

Ants display complete or holometabolous metamorphosis, consisting of four life stages: egg, larvae, pupae, and adult. The aforementioned first three stages are collectively called brood. The ant colony is almost exclusively female until the time of the nuptial flight. The life cycle of the ant colony can be divided into three parts [6]. The founding stage is initiated with the nuptial flight. The virgin queen (virgin reproductive) leaves the nest, leaving behind the queen and her sisters, who are sterile workers or virgin reproductives. When she meets one or more males and is inseminated, she finds a suitable nest site in the soil or plants and builds a first nest cell and rears the first brood of workers. The workers eventually take on the tasks of foraging, nest enlargement, and brood care, so that the queen can confine her duties to egg laying. The ergonomic stage is defined as increase in work devoted to colony growth. The reproductive stage is set after one or more seasons when the colony begins to produce new queens and males and these sexual forms go forth to begin new colonies [6].

### 1.5. Mutualistic interactions

Among ant species, there is a broad range of interesting behaviors displayed. Many species have mutualisms (interactions leading to net fitness benefits for all partners involved) [7] with other insects and plants. In one example, ants will forego predation of, and offer protection to, honeydew producing aphids and mealybugs, as the ants are attracted to this sugar water sap as a food source [8]. In some ant-plant interactions, ant transport and disperse plant seeds (myrmecochory) with elaiosomes or food bodies rich in lipids, amino acids, or other nutrients that are attractive to ants, collectively known as a diaspore. Once the diaspore is carried back to the ant colony, the elaiosome is consumed and the seed is ejected from the nest [9–11]. A well-studied symbiotic relationship with plants has been reported between *Acacia* trees and the several ants in the genus *Pseudomyrmex*. The acacia tree produces thorns, which are used as nesting sites for the ants, and Beltian bodies, used by the ants as a food source. These ants offer protection to the trees from herbivorous arthropods and vertebrates and destroy competing plants trying to establish themselves, nearby [7]. Leaf cutter ants (genera, *Atta* and *Acromyrmex*) are notorious for cutting pieces of leaves or flowers and carrying them back to their nest to use as a suitable nutritional substrate for cultivating fungal growth to help degrade cellulose and other plant products inaccessible to the ants.

### 1.6. Parasitic relationships

In contrast to symbiotic interactions that ants have with plants and other insects, various species have parasitic relationships among each other. One such case is slavery or dulosis. In the genus *Polyergus*, workers steal larvae and pupae from the genus *Formica*. These enslaved immature insects develop into adult workers and to carry out colony maintenance tasks for their captors [12]. Other examples of parasitism occur in parasitized host colonies lacking a worker caste system in the fire ant genus, *Solenopsis*, and *Teleutomyrmex*. Here, the parasitic ant *Solenopsis daguerrei* lacks a worker caste system, so all the adults are reproductive males and females. Parasitic queens attach to the host queens and fire ant workers care for them in a similar manner to their own mother queen, whereas the parasite brood is reared by the host workers simultaneously with the host brood [13, 14]. These parasites are permanent and spend their entire life cycle in the nest of the host species. Most often, the parasite inhibits egg production of the mother queen, causing an eventual collapse of the colony [15].

### 1.7. Predatory behavior

Many ant species exhibit an extremely predatory behavior. In the Afrotropical region, in the genus, *Dorylus*, the African driver ants, also known as army or legionary ants, have colonies with millions of inhabitants. They have no permanent nest structure, as they move their nesting sites regularly and forage for food in large swarming columns or groups, preying mainly on insects, arachnids, and earthworms [16]. In another example, extraordinary cooperative behavior is exhibited during nest construction by the weaver arboreal ants, in the genus *Oecophylla*, living in Afrotropical regions. These ants link their bodies together to form chains by grasping the petiole of an adjacent worker with their jaws. The living chains are used to position leaves together. Silk-producing larvae contribute their secretions to allow the leaves to be held together, eventually forming a tent and eventually a nest [17].

### 1.8. Pollinators

While bees, flies, and hummingbirds are thought as agents of plant pollination, some ant species have been found to serve as effective pollinators. This is despite observations, in general, that antibiotic compounds produced by the metapleural and poison glands of most ants tend to suppress pollen germination and pollen-tube growth. Worker ants, *Proformica longiseta* have been found to serve as pollinators of a mass flowering wood plant, *Hormathophylla spinosa,* in the high mountain area of the Sierra Nevada mountains. The ants were found to make contact with the plant reproductive organs when foraging for nectar and were found to transfer large numbers of pollen grains, contributing to increasing the number of viable seeds [18].

## Acknowledgements

This work was supported in part by grants from the National Institutes of Health National (GM058264) and the National Science Foundation (1626326).

## Conflict of interest

The author declares that there is no conflict of interests regarding the publication of this chapter.

## Author details

Vonnie D.C. Shields

Address all correspondence to: vshields@towson.edu

Biological Sciences Department, Fisher College of Science and Mathematics, Towson University, Towson, MD, USA

## References

[1] Holldöbler B, Wilson EO. The Ants. Cambridge, MA: Belknap/Harvard University Press; 1990. 733 pp

[2] Wilson EO, Carpenter FM, Brown WL. The first Mesozoic ants. Science. 1967;**157**:1038-1040

[3] Wilson EO, Carpenter FM, Brown WL. The first Mesozoic ants, with the description of a new subfamily. Psyche. 1967;**74**(1):1-19

[4] Bolton B. Identification Guide to the Ant Genera of the World. Cambridge, Massachusetts: Harvard University Press; 1994. 222 pp

[5] Wang Q, Goodger JQD, Woodrow IE, Elgar MA. Location-specific cuticular hydrocarbon signals in a social insect. Proceedings of the Royal Society of London B. 2016;**283**:20160310. DOI: 10.1098/rspb.2016.0310

[6] Oster GF, Wilson EO. Caste and Ecology in the Social Insects. NJ: Princeton University Press; 1978. 372 pp

[7] Clement LW, Köppen SCW, Brand WA, Heil M. Strategies of a parasite of the ant Acacia mutualism. Behavioral Ecology and Sociobiology. 2008;**62**:953-962. DOI: 10.1007/s00265-007-0520-1

[8] Heil M, Rattke J, Boland W. Post-secretory hydrolysis of nectar sucrose and specialization in ant/plant mutualism. Science. 2005;**308**:560-563. DOI: 10.1126/science.1107536

[9] Heil M, McKey D. Protective ant–plant interactions as model systems in ecological and evolutionary research. Annual Review of Ecology, Evolution, and Systematics. 2003;**34**: 425-453. DOI: 10.1146/annurev.ecolsys.34.011802.132410

[10] Heil M, Baumann B, Krüger R, Linsenmair KE. Main nutrient compounds in food bodies of Mexican Acacia ant-plants. Chemoecology. 2004;**14**:45-52. DOI: 10.1007/s00049-003-0257-x

[11] Heil M. Ant-Plant Mutualisms. Wiley Online Library; 2010. DOI: 10.1002/9780470015902.a0022558

[12] Ito F, Higashi S. Temporary social parasitism in the enslaving ant species *Formica sanguinea* Latreille: An important discovery related to the evolution of dulosis in *Formica Ants*. Journal of Ethology. 1990;**8**:33-35

[13] Bruch C. Notas preliminares acerca de *Labauchena daguerrei* Santschi. Revista de la Sociedad Entomológica Argentina. 1930;**3**:73-82

[14] Calcaterra LA, Briano JA, Williams DF. Florida Entomologist. 2000;**83**:363-365

[15] Silveira-Guido A, Carbonell J, Crisci C. Animals associated with the *Solenopsis* (fire ants) complex, with special reference to Labauchena daguerrei. Proceedings Tall Timbers Conference on Ecological Animal Control by Habitat Management. 1973;**4**:41-52

[16] Gotwald WH Jr. Predatory behavior and food preferences of driver ants in selected African habitats. Annals of the Entomological Society of America. 1974;**67**:877-886

[17] Devarajan K. The antsy social network: Determinants of nest structure and arrangement in Asian weaver ants. PLoS. 2016;**11**(6):e0156681. DOI: 10.1371/journal.pone.0156681

[18] Gómez JM, Zamora R. Pollination by ants: Consequences of the quantitative effects on a mutualistic system. Oecologia. 1992;**91**:410-418. DOI: 10.1007/BF00317631

Chapter 2

# Structure, Diversity and Adaptive Traits of Seasonal Cycles and Strategies in Ants

Elena B. Lopatina

Additional information is available at the end of the chapter

http://dx.doi.org/10.5772/intechopen.75388

**Abstract**

This chapter is a review of data on structure, diversity and adaptive properties of seasonal cycles in ants. Most tropical ants demonstrate homodynamic development. They do not show any developmental delays and all-year round the ontogenetic stages from egg to pupa exist in their nests. Some of the quasi-heterodynamic species have permeated into the regions with warm temperate climate but a true diapause did not evolve. Most temperate and all boreal climate ants are true heterodynamic. They manifest a real winter diapause (prospective dormancy) in their annual cycle. Thus, a variety of forms of dormancy, which were found in ants, extend from elementary quiescence to deep diapause. Heterodynamic ants use two main seasonal strategies with respect to brood rearing: strategy of concentrated brood rearing (*Formica* type) and strategy of prolonged brood rearing (*Aphaenogaster* type, *Myrmica* type). The larval stages at which diapause can occur are extremely variable among ants. The evolution of seasonal life cycles and possible ways of origin of diapause in ants are discussed. The subtropical (quasi-heterodynamic) and tropical (preadaptational) evolutionary paths to true heterodynamic development are considered. It is concluded that similar seasonal adaptations could arise in the evolution of ants independently many times and usually are not tightly bound to the taxonomic position of species.

**Keywords:** Formicidae, ants, seasonal development, life cycles, homodynamic, heterodynamic, diapause, exogenous, endogenous, temperature, climate, evolution

## 1. Introduction

The life of ants as ectothermic organisms is closely connected with the seasonal variations of ecological factors, such as temperature, rainfall, humidity, availability of food, etc., occurring

during the year. Certain climatic changes usually manifest even in tropics and cause the corresponding modifications in behavior and development of ants and other insects, but they are much more pronounced on the north. In the regions with temperate and boreal climate, the annual changes of climatic conditions possess a hard impact to the life cycles of all organisms living there. In different ant species, various annual cycles of behavior arise. They have been described in many reviews (for example, see [1, 2]). However, the seasonality of the behavioral cycle in ant colonies is strictly subordinated to the seasonality in developmental processes such as oviposition or larval rearing. Each ant species demonstrates certain annual developmental cycle because the processes of the colony development cannot proceed without interruption during the whole year.

To optimize the survival and growth of ant colonies, the entire warm season should be used for larvae rearing and for production of the greatest quantity of new adults. That is why the brood development should start in spring as earlier as feasible and prolong as longer as possible in late summer. At the same time, only the brood stages that are able to survive during the winter should occur in ant colony by the end of autumn. To ensure this, the special physiological mechanisms evolved which provide synchronization of the colony development with the yearly climatic periodicity.

The first investigations devoted to seasonality in the development of ants were performed by Yozhikov who studied the phenology of the development in several ant species from Central Russia [3]; by Headly who characterized the seasonal development of two *Leptothorax* species and *Aphaenogaster fulva aquia* in North America [4, 5]; by Talbot who carried out the phenological studies on two North American *Myrmica* species [6]; and by Eidmann who described the hibernation in ants and mentioned that the *Formica* species overwinter without a brood [7].

Despite extensive studies of arthropod dormancy and seasonality, myrmecologists paid very little attention to the role of seasonality in ant ecology, as well as in the evolution of the life cycles of these social insects. Literature on this topic is not rich. Quite a few publications specifically devoted to seasonal development and the phenology of ants. As a rule, such data can be found in the investigations dealing with the biology and ecology of certain species. Even less frequently, the subject of the study was the regulation of annual cycles of the ant development. Most often, these problems were affected accidentally and stood in the background and pushed back by the main aims of the study (for example, see [8]). The seasonal development of the ants *Myrmica* was studied to the greatest extent, mainly due to the works of M. Brian (for reviews, see [1, 8–10]).

This paper contains a review of literary and proprietary data on the structure, diversity, adaptive features and evolution of seasonal cycles and strategies of the seasonal development in ants. We have studied the seasonal cycles in more than 80 species belonging to more than 20 genera from different regions of Russia and the former USSR, ranging from warm temperate to cold temperate and boreal climatic zones. Our field and laboratory studies have allowed us to map the diversity of annual cycles, to reveal the underlying ecophysiological and social mechanisms of control and to develop ideas on possible pathways in the evolution of the seasonal cycle in ants.

The main study methods we have used were laboratory experiments and field phenological observations. In experimental studies, the ant colonies were divided into fragments each consisted of workers, queens and the brood. In the case of a small number of workers in the colony or for monogynous species with a single queen in the nest, entire natural colonies were used. Colony fragments (or the whole colonies) were kept in artificial plastic nests, which were randomly distributed over various experimental regimes (different photoperiodic conditions and constant temperatures or thermal periods) and experimental regimes were maintained in special thermostats. Our methods of laboratory cultivation of ants provided the opportunity to observe and to study all stages of annual cycle including the overwintering in a refrigerator under the temperatures of 3–5°C.

## 2. Peculiarities of seasonal development in ants as social insects

In social insects, in addition to the life cycles of individual individuals, there is a life cycle of the colony, as an integrated, superorganismic system. It consists of the processes of the development of individuals, but it is not equivalent to a simple summation of them. Social regulation mechanisms arise in the evolution and are realized through interactions between members of the colony. They control the physiological state and the development of individuals, depending on the ecological situation and colony needs. Such "collective regulation" of development is absent in solitary species and largely determines the specificity of seasonal development in ants [11–16].

Colonies of ants are not only perennial but usually have unlimited life cycle under favorable conditions of environment [17]. Not only queens but even workers can survive for several years. In monogynous species, all workers and the brood are the offspring of the only queen. So while the queen is alive, all the population of a colony can relate to one and the same genetic generation during consistent years. In polygynous species with several queens in a nest, all individuals inhabiting a single colony may pertain to various generations which overlay each other. However, the seasonal life cycle of the colony is not associated with differences arise between generations. It embraces the regular seasonal variations in physiological state of all individuals in a colony which entail the orderly changes in behavioral and developmental patterns. Therefore, we determined the seasonal life cycle of an ant colony as the annual cycle of physiology, behavior and development [18].

After the spring awakening, the queen commences laying eggs, the development of larvae begins and pupae appear. There are workers and reproductive individuals among new adults. Oviposition and brood development continue throughout the warm period of the year and cease in the autumn, when insects begin to prepare for wintering, go to special shelters and spend the winter in inactive state. The annual cycle of colony development is a collective and highly organized process and includes the individual development of immature stages (the brood) and regular seasonal changes in the physiological state and reproductive activity of adults (workers and queens).That is why the growth and development and the beginning and termination of diapause in ants can be considered and studied both in individual level and at the level of the entire colony. And this is not the same thing, since all these processes are under the control of the mechanisms of social regulation and integrated reactions of the colony to the changes in environmental conditions. In this connection, diapause of ant larvae can be (and usually happens) facultative at the level of an individual, but at the same time obligatory at a colony level (in endogenous-heterodynamic species).

## 3. Homodynamic type of seasonal development

On the general background of an insufficient study of the phenology in ants, the tropical regions of the Earth are especially prominent. Analysis of meager data shows that at any time of the year, in the nests of most tropical species studied, all the stages of ontogenesis from the egg to the pupa are present, and development retardations are absent. Such continuous all-year round development without the obligatory onset of periods of physiological dormancy we call homodynamic, using the terminology of E. Roubaud [18–21].

However, homodynamic species often have a certain seasonal structure of the annual cycle: on a general background of continuous development, there may be significant seasonal fluctuations in the number of certain ontogenetic stages, as well as seasonal association of the rearing of alates and nuptial flights. Thus, in *Cataulacus guineensis* from Ghana, the brood population has two maxima—in May and in September, the alate females and males are numerous in the nests in July–October, and the rest of the time there are few or none alates at all [22]. The larvae of alate reproductives of *Camponotus sericeus* in India develop from October to July, and alate females and males can be found in nests all-year round, but their nuptial flight occurs in September–October [23, 24]. Colonies of the tropical ant *Paltothyreus tarsatus* (Republic of Côte d'Ivoire) grow alate females in spring, and males in autumn; reproductives live in maternal nests for a long time and leave for mating only in February [25]. In definite periods of the year, alates of *Anoplolepis longipes* are grown up in Papua New Guinea [26] and in the Seychelles [27] and alates of *Camponotus detritus* in Namibia [28]. The factors that cause such seasonality in tropical species are not yet clear enough.

In addition, periodicity of oviposition was noted for a number of species. For example, in the natural colonies of *Anoplolepis longipes, Cataulacus guineensis* and species of the genus *Rhytidoponera* [29], there are two egg abundance maxima throughout the year. Rhythm of oviposition was also found in a tropical species *Linepithema humile*, widespread in Southern France, both in natural and laboratory conditions [30]. The periodicity of egg laying is also a characteristic feature of the nomadic ants of the tropical subfamily Dorylinae and especially of the Neotropical species. In all species, the cyclic brood rearing is clearly associated with behavioral cycles: oviposition occurs during stationary phase of reproductive cycle that alternates with nomadic phase, during which the ants feed the larvae that emerge from the eggs [31–33].

The homodynamic development of some tropical ants was observed in the laboratory. G. Terron maintained colonies of the African species *Tetraponera anthracina* at 25–26°C for several years and reported that development of the brood occurred continuously and reproductives periodically appeared [34]. But he observed alternating periods of egg laying and reproductive dormancy of queens. The ant *Monomorium pharaonis*, a widespread synanthropic species, which was imported from tropical Africa to Europe and North America, now inhabits many heated buildings. It was found that in the nests of this species under optimum conditions (27–30°C, relative to 60–65%), brood developed continuously with cyclic rearing of alate reproductives at regular intervals [35–41]. We also kept several colonies of *M. pharaonis* found in St. Petersburg in the laboratory for 2 years and revealed that at any temperatures above the threshold, which is 17.7–17.8°C in this species [42], brood development occurred without any delays. However, at a near-threshold temperature of 17°C, the mortality rate of all brood categories, as well as of adult ants, sharply increased, and the colonies died within a month. Thus, this tropical species cannot tolerate even short periods of temperature decrease, which confirmed by the available data on its habitation in Europe only in heated buildings [43].

We observed homodynamic development in two species from tropics: *Pheidole sexspinosa* from the Tonga archipelago and the *Tetramorium semillimum* from the Seychelles. We kept them in the laboratory for more than a year, varying the temperature and photoperiod within acceptable values, i.e. within those that can be observed in the natural habitat of these species (18–25°C, 10–16 hours of light per day). All this time, there was a continuous development in the colonies.

It is well known that typically tropical insects cannot survive for a long time at temperatures well below the optimum and, especially, below the developmental threshold [44]. Therefore, it can be argued that the tropical ants in their majority are not adapted to survive during the cold periods of the year. However, some ant species can demonstrate continuous all-year round development even in subtropical environments. For example, in the central regions of Texas (USA), *Pseudomyrmex* sp. occupies the stems of a mimosa, and all ontogenetic stages are present year round, but in different amounts: eggs and larvae are numerous in winter and pupae in spring and summer [45]. Thus, we can assume that the development and pupation of larvae of this species continues in winter, but much more slowly than in summer. This decrease of brood developmental rate leads to a significant decline in the number of pupae in wintering nests. According to Rissing, new workers of *Messor* (*Veromessor*) *pergandei* in Arizona (USA) appear from pupae also throughout the year [46].

## 4. Heterodynamic cycles of development

Annual developmental cycles of ant colonies, in which a diapause arises naturally, we call after Roubaud heterodynamic [18–21]. They have a distinct seasonal structure: the period of diapause (the phase of dormancy) is regularly replaced by the period of development (the active phase of the cycle), after which a new period of diapause occurs, and so on. Generally, the phase of dormancy in the annual cycle coincides with the period of unfavorable climatic and (or) food conditions. It is characterized by a lack of larval development and, as a rule, of queen oviposition and the presence in the nests of only certain (usually diapausing) brood categories. During the active phase of the annual cycle, eggs are laid and the larvae are reared. Ants grow up new workers, as well as alate females and males.

### 4.1. Quasi-heterodynamic development

Some ant species recently penetrated from tropics or subtropics to areas with a warm temperate climate and successfully settled there. Two species of fire ants, *Solenopsis richteri* and

*S. invicta,* were imported to the United States from South America: the first one, around 1920, and the second one, in the early 1940s [47, 48]. *S. invicta* significantly expanded its original range fairly far north and is now prevalent in most southeastern states [48, 49]. The distribution of *S. richteri* is limited mainly to the northern parts of Alabama and Mississippi and the southern part of Tennessee [50].

Several authors have shown that in southern United States (Mississippi, Florida, South Carolina, etc.) both fire ant species remain essentially homodynamic: eggs, larvae and pupae of workers are present in their nests all-year round [48, 51, 52]. However, the number of immature stages varies considerably during the year, and in winter, it is very small. In February–March, the number of eggs increases sharply and new larvae develop and begin to pupate in April (workers) and May (alates). At this time, the quantity of brood categories is maximal and reaches 40–45% of the entire biomass of the colony. Reproductive individuals appear from pupae in June and fly out of the nests at least five times during the summer. The second small peak of the brood population (up to 35% of the biomass) in the nests coincides with the first cooling in September–October. At this time, all larvae and pupae belong only to the caste of workers. In November–December, the number of larvae and pupae sharply decreases and reaches a minimum (less than 2% of the biomass of the colony) in January. Thus, the number of immature stages in fire ants directly depends on the environmental temperature.

It has been determined that in the most northern populations of *S. invicta,* oviposition and brood development are suspended in the coldest months, i.e. eggs and pupae are absent in the nests in winter [48, 53]. It can be assumed that there is no diapause of larvae in fire ants and larval development ceases only when the ambient temperature falls below the developmental threshold, which is about 17°C according to laboratory experiments [54]. We can also assume that many eggs, larvae and pupae (as well as adult ants) die during the winter in northern populations of this species. The winter mortality of *S. invicta* workers was noted in several works (for example, [55]). Nevertheless, the colonies successfully survive in these conditions. Consequently, these ants already have some physiological adaptations that allow them to tolerate a sufficiently long stay at low positive temperatures.

Such tropical species, adapted to live in regions with cold winters without forming real diapause, we call quasi-heterodynamic [18]. They are characterized by the potential for unlimitedly long development under favorable conditions inherent in homodynamic species. The development of their brood ceases only at temperature below the developmental threshold and ants spend the winter in a quiescent (cold coma) state suffering from more or less strong mortality, but in general the colonies overwinter successfully.

The Argentine ant *Linepithema humile* demonstrates another example of quasi-heterodynamic development. This tropical species was imported from South America and now widespread in many parts of the world. Its phenology, investigated in the USA [56] and in Europe [30], is very similar to the seasonal development of fire ants. In the southern part of California, only a few larvae of workers and very few eggs can be found in overwintering nests of this species. At this time of year, more than 90% of the colony biomass is made up by adult ants. The new seasonal cycle begins in late February–early March, when the queen starts to lay eggs. The larvae hibernated in the nests complete their development by the middle of March. In summer, brood makes up about 50% of the colony biomass, but in October its number decreases sharply and gradually reaches its minimum by December. On the southern coast of France, the Argentine ant has similar phenology. Most small larvae hibernate in the nests, but in small numbers, and there are also some medium and large larvae and very rarely eggs. Renewal of development is observed in March.

Since the fire ants and Argentine ants recently permeated to the areas with cold winters, we tend to think that they do not yet have diapause. However, it has been experimentally determined that only the overwintered colonies of the Argentine ant can grow up a lot of alate females, which appear in the nests in the south of France at the beginning of the summer season [57]. This suggests that there are seasonal changes in the physiological state of colonies that are similar to diapause.

According to the literature date, many subtropical ants do not have any brood in their nests during the winter, for example, *Ponera pennsylvanica* in the state Missouri [58], *Pogonomyrmex rugosus* and *P. subnitidus* in Southern California [59] and *Prenolepis imparis* in the states Missouri, Ohio and Florida [60–62]. However, without experimental laboratory studies, it is impossible to conclude whether all of them are quasi-heterodynamic and overwinter without brood due to its death, or, conversely, they have a stable winter diapause of the queens, i.e. belong to the group of true heterodynamic species.

We discovered and investigated quasi-heterodynamic developmental cycles in several ant species living in regions with subtropical and warm temperate climates. In the colonies of *Tetramorium nipponense* and *Pachycondyla chinensis*, which were collected in the south of Japan and were kept in the laboratory at temperature of 25–27°C and photoperiod of 16 h of light per day, the oviposition of queens and the development and pupation of larvae did not cease during the year. After a gradual decrease in temperature, the colony successfully overwintered at 7–8°C. At the same time, however, all brood died, which probably occurs during the overwintering in natural conditions of the south of Japan.

In the experiments on *Monomorium kusnezovi* from Turkmenistan, the growth, development and pupation of larvae continued uninterruptedly at temperatures exceeding the developmental threshold of 20°C for eggs and 21.5°C for larvae and prepupae [42]. Thus, larvae of this species do not have diapause, which is a sign of quasi-heterodynamic seasonal cycle. At 20°C, the larvae ceased growth, but the oviposition did not stop, and by the beginning of overwintering, eggs and larvae of all ages were still present in the nests. However, eggs, as a rule, did not survive during hibernation in the laboratory, but the larvae hibernated more successfully. We assume that, in natural conditions, these ants go on hibernation with eggs and larvae, but eggs and part of the larvae die during the winter. In the middle of April in the Central Kopet Dag when we dug out the nests of *M. kusnezovi*, we always found small packets with eggs and larvae of younger instars and a small number of the third instar larvae lying separately, which already began to pupate at this time of a year. Obviously, these larvae overwintered in the nests but the eggs could appear in the early spring. Thus, it seems likely, though not completely proved, that a certain number of eggs can be preserved in the nests of *M. kusnezovi* until spring.

Two *Pheidole* species (*P. pallidula* and *P. fervida*) were collected in Turkmenistan and Russia, where they inhabit the regions with temperate climate. They also proved to be quasi-heterodynamic. *P. pallidula* is common in southern Europe, the Caucasus and Central Asia [63]; the distribution area of *P. fervida* includes Southeast Asia, Japan, southern Kurils and the south of Primorye [64]. In Primorye, *P. fervida* probably survived from the Tertiary period, when the climate there was much milder. In our experiments with both species, the queens continued to lay eggs and the larvae developed and pupated at any temperatures exceeding the developmental threshold (about 18°C for *P. pallidula* [42]). Any forms of diapause were absent. Under optimum conditions at 25–28°C, we observed continuous development for two or more years.

For *P. pallidula* from Turkmenistan, we found that when the temperature decreased to 20° C, the queen's productivity declined significantly, but queens did not stop laying eggs, and the larvae developed and pupated. At a temperature of 17°C, which is only slightly below the developmental threshold for eggs and larvae, oviposition continued, but the eggs did not develop. Pupae and prepupae were the first to die, and then all brood stages gradually perished. Consequently, it could be assumed that in nature all brood in the nests of this species die even before the beginning of winter. Our field studies in Turkmenistan have shown that there was no brood in the nests of *P. pallidula* in early spring indeed. It appeared later from the eggs laid by queens. This is confirmed by the observations made by Passera [65–67] in the south of France.

*P. fervida* is somewhat better adapted to the temperate climate in the south of Primorye and its brood probably dies out only partially during overwintering. In our experiments with autumn colonies of this species, even at a temperature of 17°C and under short day conditions (10 hours of light per day), the queen continued oviposition, and larvae still pupated. We maintained one colony, in which there were originally more than a hundred larvae of different size, under these conditions during 7 months. Over this time, some of the larvae perished, but more than half of them pupated. This situation was strikingly different from the one we observed in similar experiments with truly heterodynamic ants: in most cases at 17°C, the development very quickly ceased due to the onset of diapause.

When we decreased the keeping temperature for the autumn colonies of *P. fervida* up to 10–12°C, the workers began to dismember and to throw out the prepupae and pupae from the nests and gradually destroyed them all. There were only eggs and larvae of all instars in the nests. However, part of the eggs and larvae of the first and second instar perished during the hibernation in the refrigerator at 3–5°C, while the older larvae overwintered more successfully. It would be extremely interesting to compare the population of the colonies of this species in late autumn and early spring to assess the ability of the larvae to survive during the winter in natural habitats. Probably, the similar regulation of the seasonal developmental cycle exists in *Pheidole morrisi,* which hibernates with larvae in the north of Florida [68].

### 4.2. True heterodynamic development

Most temperate and all boreal climate ants are true heterodynamic. They possess real winter diapause in their annual cycles (prospective dormancy) [18]. The presence of this diapause provides a more successful wintering by increasing the physiological resistance of larvae and adult ants to unfavorable winter conditions. In the literature, there is practically no data on the tolerance of developing and diapausing larvae, other developmental stages and adult ants to low temperature and other unfavorable environmental factors. Plateaux [69] noted that *Temnothorax nylanderi* in an active physiological state could not successfully overwinter:

when he put summer colonies of this species into the refrigerator, eggs, pre-pupae and unpigmented pupae quickly died and began to rot, causing the death of the entire colony. Similar results were obtained in our experiments with colonies of *Lasius niger, Lepisiota semenovi, Myrmica rubra, M. ruginodis* and *Plagiolepis compressus,* which we put in the refrigerator at a temperature of about 5°C in summer. Eggs, pre-pupae and pupae died within 1–3 weeks, but larvae remained alive.

True heterodynamic seasonal cycles occur in the vast majority of ant species living not only in temperate and cold climates, but also in subtropics and even in tropics. The presence of long-term developmental delays was noted, for example, in all five species from the *Rhytidoponera impressa* group, widespread in the forests of Eastern Australia. These species demonstrated a strict seasonality of development: during the winter months, only small and medium-sized larvae and very rarely eggs were found in their nests. This situation was observed both in subtropical and tropical regions of Australia [29].

It is clear that for the occurrence of heterodynamic development in tropics, any seasonal changes in environmental conditions have to exist. Since the annual rhythm of the climate is usually quite distinct in the tropical regions, and its absence, on the contrary, is very rare situation, heterodynamic seasonal cycles should probably be widespread in tropical ants. The hibernation and diapause are widespread in tropical insects, but mechanisms of the regulation of heterodynamic cycles in tropics are far from understanding and explaining yet [70].

It should be assumed that heterodynamic development is more common for ants in subtropics, because the seasonal rhythm of the climate there is much more pronounced than in the tropical zone. Indeed, most of the subtropical species studied demonstrate the cessation of development in winter. Some of them do not have brood, while others overwinter with larvae: *Camponotus kiusiuensis* [71], *C. nawai* [72] and *Leptanilla japonica* [73] in southern Japan, *Temnothorax monjanzei* in Algiers [74] and *T. melas* on the Corsica island [75], *Tapinoma minutum* in New South Wales [76], *Polyrhachis vicina* in southern China [77, 78] and *Pheidole morrisi* [68] and *Dolichoderus mariae* [79] in the north of Florida. However, it is impossible to decide without special experiments whether these subtropical ants possess a real diapause, or they are quasi-heterodynamic and stop their development during the cold season due to direct influence of low temperature.

## 5. Two seasonal strategies of brood rearing in heterodynamic ants

Analyzing the structural diversity of heterodynamic seasonal cycles in ants, we identified two fundamentally different directions in their evolution, and accordingly, two seasonal strategies for brood rearing [18].

### 5.1. The strategy of prolonged brood rearing

The ants are more likely to follow the strategy of prolonged brood rearing. This strategy is based on the ability of larvae to enter into a diapause and to continue development over the next summer (**Figure 1**). Depending on the composition of the overwintering brood and the stage at which diapause is observed; we distinguish two structural types of developmental cycles [80].

*Aphaenogaster type.* Larvae fall into a diapause at the end of summer and all the remaining pupae manage to complete the development before the onset of colds, but the queens have no diapause and do not cease oviposition until the late autumn. Therefore, not only diapause larvae but also eggs and young larvae overwinter and survive, at least partially, during the winter. Thus, the formation of the wintering population of the colony is determined by the appearance of larval diapause in species with this type of seasonal cycle. These species are apparently limited in their distribution by subtropics and the southernmost regions of the temperate zone.

*Myrmica type.* The diapause starts both in larvae and queens at the end of summer. Therefore, before the winter comes, larvae emerge from the laid eggs, and all pupae develop into adults. The overwintering brood is represented only by larvae. This annual cycles are typical for most ant species living in a temperate climate zone. The induction of diapause in both larvae and queens plays an equally important role in the synchronization of *Myrmica* type cycles with the annual rhythm of the climate and in the formation of the wintering composition of the colony.

The annual cycles of ants that overwinter with brood have the most complex seasonal structure (**Figure 1**). All hibernating larvae usually pupate during the summer season. They give the first peak in the number of pupae in the nests. As a rule, alate females and males develop from most of them. Some of the larvae that emerged from the eggs which were laid in spring and early summer can pupate during the same growing season. This is the so-called a rapid or summer brood, according to Brian [81, 82]. It develops without diapause and gives the second peak in the number of pupae. All other larvae that emerge from the eggs within the season, fall into a diapause, hibernate and finish their development only next summer. This

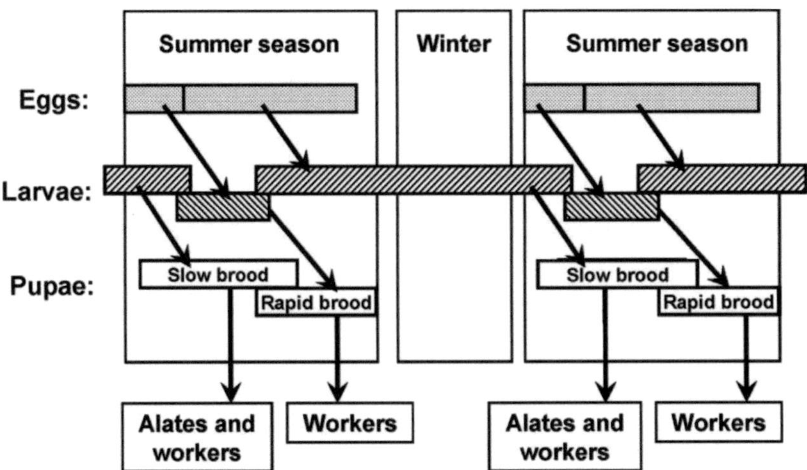


**Figure 1.** The strategy of prolonged brood rearing. Further explanation in the text.

is so-called a slow or winter brood [81, 82]. Thus, two complete cycles of brood development from the egg to imago take place in a colony during each year (**Figure 1**): a summer cycle that begins and ends within one growing season, and a winter cycle, in which larval development is interrupted by their diapause.

Rapid brood is found in most species from the temperate zone. It may be absent in species and populations from the northern regions, where summer is short, as well as in species with very slow individual development. For example, the rapid brood is absent in *Amblyopone pallipes* from Massachusetts [83]. It can be assumed that this species is characterized by very slow development, although there are no data on this problem. The Japanese ant *Leptanilla japonica* has an extremely specialized annual cycle without rapid brood [73]. The queen lays the only portion of eggs at the end of July to the beginning of August. The larvae develop, overwinter at the last instar and pupate in July of the following year. They develop very synchronously, so that at each moment there is only one ontogenetic stage in the nest. Adults emerge from pupae at a time when larvae appear from the next portion of eggs. Thus, in spite of the fact that this species lives in subtropical climate, only one cycle of brood development occurs during a year.

The strategy of prolonged brood rearing is extensive inherently, as it is realized by stretching of the development of individuals for two or more summer seasons. The appearance of larval ability to fall into a diapause gave the ants a unique way for adaptation to the life in a temperate climate and especially in high latitudes. Therefore, the strategy of prolonged development is most common among the ants living there. It has a number of adaptive advantages.

#### *5.1.1. More complete use of a favorable period for the development and available thermal resources*

Since the larvae are always in the nest, the workers can feed them from early spring to late autumn. Immediately after the end of the winter, as soon as it becomes a little warmer, the ants carry larvae from the underground chambers to the upper, warmed by the sun, horizons of the nest (beginning with the largest larvae), creating the best opportunity for larval growth and development. As it becomes warmer and the amount of available food increases, the ants carry more and more small larvae to the upper levels of the nest and begin to feed them. According to Peakin for *Lasius flavus*, the overwintered larvae of the first and second instars remain in the deep and cold nest chambers for the longest time and therefore complete their development only by the end of the summer [84]. Just the same was recorded for *Temnothorax nylanderi* by Plateaux [69, 85]. In the spring, the workers of this species also start with the feeding of the largest overwintered larvae which develop into alates.

The autumn period of larval rearing is also of great importance for most ants with a wintering brood. For example, in the central part of European Russia in the colonies of *M. rubra*, the pupation of larvae ceases, as a rule, in the first half of August due to the onset of larval diapause. However, until the middle of September, the larvae continue to hatch from remaining eggs, to grow and develop. New third instar larvae appear and soon fall into a diapause. Moreover, in *Myrmica* diapausing, third instar larvae retain their ability to feed and to grow slowly (without the development of imaginal buds), and therefore, they continue to gain the weight in the autumn period [9, 86]. The same autumn growth of the last instar larvae was observed in *Leptothorax* species [69]. According to our data, larval growth in diapause state is a characteristic feature primarily for species in which larvae enter into a diapause and overwinter in the last or all instars, *Aphaenogaster, Lasius, Leptothorax, Manica, Messor, Myrmica, Solenopsis, Temnothorax* and *Tetramorium* and a number of species in genera *Camponotus, Crematogaster* and *Monomorium.*

Thus, in the species using the strategy of prolonged brood rearing, workers are engaged in feeding of larvae until the final onset of cold weather and give them all the surplus food produced during this period (minus the amount of nutrients that the workers accumulate in their fat body). This makes it possible to maximize the total mass of the wintering brood and, consequently, to grow up earlier the first workers, as well as reproductives, next spring. In addition, the biomass accumulated by larvae is also a reserve of nutrients for the colony: in the case of food shortage, ants can eat a part of the brood (mainly eggs and small larvae) in order to survive and to feed the largest larvae [2, 87].

#### *5.1.2. The ability to adapt to the duration of the warm period of a year by changing the amount of the rapid brood*

Such adaptation path can be realized during the penetration of the ants into more northern areas [88] and in connection with the local variability of climatic conditions from year to year. In the north, where summer is short and heat resources are limited, the ants can grow up much less number of the rapid brood than in the south. For example, in the south of France, *Temnothorax unifasciatus* has numerous rapid brood, and in cooler Belgium, only small amount of it [89]. According to our data for *Lasius niger, L. flavus, Myrmica rubra, M. ruginodis* and *M. scabrinodis* from the central part of European Russia, all overwintered larvae pupate in the spring and in the first half of summer, and then a numerous rapid brood larvae also pupate. Simultaneously, at the latitude of St. Petersburg, where the warm resources available to the ants of these species are much less, usually only a small part of larvae emerging from the eggs pupate during the same summer.

In the far north, where the summer is even shorter and the warmth is even less than in St. Petersburg, the ants generally never have rapid brood [88]. This was demonstrated in our studies on *Leptothorax acervorum* and *Myrmica kamtschatica* from the upper reaches of the Kolyma River and *M. rubra* and *M. ruginodis* from the coast of the White Sea near the Arctic Circle. Similar changes in the amount of rapid brood occur when the average summer temperature increases or decreases from year to year. For example, in cool summer, *Temnothorax nylanderi* may have no rapid brood, although it is usually quite numerous [69]. We observed the same in St. Petersburg region for *Lasius niger, Leptothorax acervorum, Myrmica rubra, M. ruginodis* and *M. scabrinodis,* in colonies of which there was no rapid brood in cool years, and even some larvae could stay for repeated overwintering.

#### *5.1.3. The ability to stretch the development of larvae for two or even three summer seasons*

Lack of thermal and/or nutritional resources during the summer is not uncommon situation in areas with cold temperate climate. As a result, some overwintered larvae that do not reach the size sufficient for pupation, fall into a diapause repeatedly and hibernate the second time. Repeated overwintering of some larvae was noted for *Camponotus aethiops* [90], *Temnothorax nylanderi* [69], *Leptothorax acervorum* [91], *T. grouvellei* [92] and *Myrmica rubra* [93]. We observed this phenomenon in our experiments with *Aphaenogaster sinensis, Camponotus herculeanus, C. japonicus, C. aethiops, Lasius niger, L. flavus, Leptothorax acervorum, Manica rubida, Myrmica rubra* and *M. ruginodis*. The possibility of repeated larval hibernation is of particular importance for ants living in the far north with an extremely short summer. According to our data for *Myrmica kamtschatica* and *Leptothorax acervorum* from the upper reaches of the Kolyma River, all wintering larvae pupate in warm years, but only a part of them if the summer is cold.

However, the number of larvae repeatedly overwintering probably cannot be significant in the colony. This is hampered by some social factors that limit possible changes in the structure of the seasonal developmental cycle when ants penetrate into more northern regions [88]. Therefore, ants never go over to the opportunistic strategy of stretching development for several years, so typical for many boreal and arctic insects [94, 95]. This feature restricts further spread of ants to high latitudes.

### 5.2. The strategy of concentrated brood rearing (*Formica* type)

This strategy presumes the obligatory completion of the development of larvae emerging from the eggs during one summer season, i.e. is typical for species that hibernate without brood (**Figure 2**). We named such annual cycles as *Formica* type [80], since they were first described for ants of this genus [7]. Diapause in larvae is absent, but it arises in queens at the end of summer long before the autumn cooling. Therefore, even the eggs laid by the latter have time to complete the development. All adults emerge from the pupae before cold weather comes on, and the ants prepare to overwinter. We can say that all brood is rapid in these ants.

After the onset of queen diapause, new eggs stop to appear and all existing brood gradually completes development. The diapause of queens should not occur too early, otherwise the period available for brood development would be actually reduced. Simultaneously, if diapause arises too late in the season, many larvae and pupae would be caught by the onset of winter and destroyed by the cold. That is why the moment when queens enter into a diapause is the most important for the *Formica* type annual cycles. From the point of view of using available heat resources, the strategy of concentrated brood rearing is less effective than the strategy of prolonged development. It can be realized in a temperate climate zone, where summer is short, only in the combination with increased brood developmental rate, i.e. by intensifying developmental processes, which becomes extremely important for northern species and populations living in areas with a particularly short summer.

Our studies have shown [42, 96] that this is really so: in *Cataglyphis*, and especially in *Formica*, individual developmental rates are significantly higher than in most species with hibernating larvae. Among the northern ants, *Formica* species have the shortest developmental times and develop almost twice as fast as *Myrmica* and *Leptothorax*. All six *Formica* species which we have studied were very similar in the duration of ontogenesis and the temperature sensitivity of the development [42]. Moreover, the development of *Formica* is much more thermal sensitive

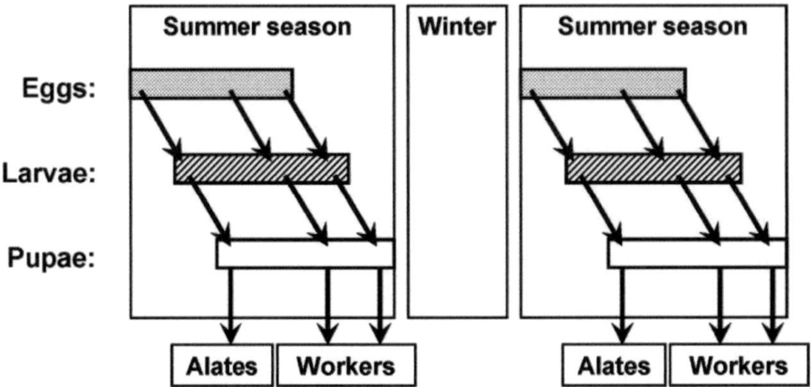


**Figure 2.** The strategy of concentrated brood rearing. Further explanation in the text.

than in *Myrmica*, due to relatively higher temperature thresholds and a higher coefficient of linear regression of the developmental rate on temperature, which allows *Formica* ants to rear brood especially intensively at higher temperatures [42, 96].

According to our data, at temperatures of 25–26°C close to optimum temperatures for *Formica* species, their developmental time from the egg to the pupa is only 20–25 days, while it is 34–35 days in *M. rubra*. Moreover, at temperatures optimal for *M. rubra* (about 22° C), their developmental time from the egg to the pupa is 40–45 days, i.e. almost twice as much as *Formica* species. Thus, in *Formica*, brood rearing really occurs more intensively. In concordance with our observations for *Serviformica* species, in European Russia, their development from the egg to the adult can be consistently realized two or three times over the summer, i.e. these ants can rear two or three large batches of the brood. In *Myrmica* species, only one full cycle of development, rapid brood, passes in the colony during the summer season. The larvae of the second cycle enter into a diapause, overwinter and finish development next summer.

Thus, it can be assumed that raised rate of ontogenesis in *Formica* species completely compensates for the shortcomings of the strategy of concentrated development and allows to significantly increase the total amount of brood reared in the temperate climate during the summer. So far, however, it is difficult to say whether such a high rate of ontogenetic processes is a special adaptation that appeared in evolution during the formation of annual cycles of *Formica* type, or the strategy of concentrated development could arise only in ants that already had a high rate of development as preadaptational.

## 6. Review of heterodynamic annual cycles in ants

### 6.1. Seasonal cycles of *Aphaenogaster* type

For the first time, the seasonal cycle of ants of the genus *Aphaenogaster* was described in the work of Headley [5], who counted the quantitative composition of 46 colonies of *A. fulva aquia* (= *rudis*) in Ohio and recorded that eggs and larvae of all sizes remained in the nests for hibernation. These data were confirmed by Talbot, who conducted research on *A. fulva* and *A. rudis* in the more southerly state of Missouri [58, 97] and much later by Mizutani and Imamura on *A. japonica* in Sapporo (Japan) [98].

We found and investigated this annual developmental cycle on *A. sinensis* (Southern Primorye, Russia), *A. gibbosa* (southwestern Turkmenistan) and *A. subterranea* (Crimea) and suggested to separate it into a particular type [80]. Apparently, it is typical for the whole genus *Aphaenogaster*. In three species studied by us, a diapause of larvae of the last (third) instar arises at the end of summer. However, queens continue to lay eggs until late autumn, thus they do not have diapause. Therefore, eggs and larvae of all three instars stay in the nests for overwintering. At least some of the eggs and junior larvae hibernate successfully, both in the laboratory and in nature. We confirmed this fact during the excavation of nests in the south of Primorye and in the Western Kopet Dag (Turkmenistan) in early spring. In accordance with our data, the same seasonal cycle is typical for *Tapinoma erraticum, T. karavaievi* and, possibly, for some *Messor* species in Turkmenistan, as well as for *Temnothorax* species, although, in all these species, eggs cannot so successfully overwinter.

Thus, of all the currently known ants from the temperate climate going on hibernation with eggs, they probably overwinter happily only in the *Aphaenogaster* species. The incapacity of the eggs for hibernation is an obvious consequence of the impossibility of diapause onset at this ontogenetic stage in ants. Our experiments with the colonies of *Lepisiota semenovi, Lasius niger, Myrmica rubra, M. ruginodis* and *Plagiolepis compressus,* which we placed in summer into a refrigerator with a temperature of 3–5°C, have shown that the eggs of these ants died at low temperatures for 2 to 3 weeks and, of course, could not survive for a fairly long winter in the natural habitat.

Despite this, in subtropics and in areas with a very warm temperate climate, where winters are mild and short, seasonal cycles of *Aphaenogaster* type, which are characterized by the absence of queen diapause and overwintering with eggs and larvae of all instars, are likely to be widespread among ants. This is confirmed by some data. Thus, living in subtropics *Linepithema humile* [56], *Polyrhachis vicina* [77, 78], *Solenopsis invicta, S. richteri, Rhytidoponera impressa* [29] and *R. metallica* [99], always overwinters with eggs and all instar larvae. The presence of eggs and larvae of all instars in the nests during the winter and even the oviposition of queens at low autumn and winter temperatures are noted for *Leptothorax niger, L. raeless* and *Temnothorax recedens* in the south of France in a very warm temperate climate, almost subtropical climate [100]. Therefore, it seems obvious that the heterodynamic annual cycles of *Aphaenogaster* type are of subtropical origin. Some species can later penetrate into areas with a warm temperate climate, while retaining the "subtropical" structure of their cycle. However, in this case, eggs perish during the overwintering period due to more severe winter conditions.

Most species of the genus *Aphaenogaster,* including those studied by Headley [5] and Talbot [58, 97], are also confined to tropics and subtropics [2]. It can be assumed that some representatives of this genus, which have spread from subtropics to areas with colder and more prolonged winter, evolved some physiological mechanisms that increased the viability of eggs at low temperatures. To ascertain their nature, special investigations are needed. Thus, according to the literature and own data, in addition to *Aphaenogaster* species [5, 58, 80, 97, 98], the same seasonal cycle is typical also for some species of *Leptothorax, Messor, Polyrhachis, Rhytidoponera, Tapinoma* and *Temnothorax*, inhabiting areas with a subtropical and warm temperate climate.

### 6.2. Seasonal cycles of *Myrmica* type

The first studies of annual cycles of such type were fulfilled in the USA by Headley on two *Leptothorax* species [4] and by Talbot on *M. schencki emeryana* [6] and on two *Temnothorax* species [101]. Then, Passera studied in detail the seasonal development of *Plagiolepis pygmaea* in Southern France [102, 103], and Sanders investigated three *Camponotus* species in the south of Canada [104, 105]. The seasonal cycles of *Messor capitatus* [106], *Camponotus vagus* [107] and *C. aethiops* [108] in the south of France, *Myrmica rubra* in Belgium [109] and *Paratrechina flavipes* in Japan [110] were studied in the same detail.

The first study, in which the annual cycle of development was observed in the laboratory, belongs to Brian [81]. He maintained two colonies of *M. rubra* and *M. ruginodis*, in artificial conditions closest to natural. Later, he investigated the growth and development in several colonies of the same species, founded by females, before the production of alates [93]. The developmental cycle of *Plagiolepis pygmaea* [103] and a number of *Leptothorax* species [69, 75, 85, 89, 92, 111, 112] were studied in much greater detail. Plateaux kept the colonies founded by the inseminated females of *Leptothorax* under conditions simulating the change of seasons, from the moment of the colony foundation until the queen death and the natural extinction of the colony (5–12 years), i.e. recorded the complete ontogeny of all colonies [75, 92, 111, 112]. Similar studies have been conducted on *Messor syriacus* [113], *M. incorruptus* [114] and *M. barbarus* [115, 116]. Based on literature data, we can include in *Myrmica* type, the seasonal developmental cycle of European species *Messor harvester* [117] and *Dolichoderus mariae* [79] from the north of Florida (USA).

The larval stages, on which the diapause can occur, are extremely variable among ants with *Myrmica* and *Aphaenogaster* types of annual cycles. Five groups can be distinguished according to the instar composition of overwintering larvae [18].

#### *6.2.1. Diapause in early (1st–3rd of 5 or 6) instars (Lepisiota, Plagiolepis, Tapinoma and some Camponotus)*

Larvae of the first three instars hibernate in *Plagiolepis pygmaea* according to Passera [102, 103]. Our data for *P. calvus, P. compressus, P. karawajewi* and *P. vladileni* are completely consistent with this conclusion. In *Tapinoma erraticum* from France, most of the larvae hibernate at the first instar and only a few at the second and third [118]. We found the same in *T. erraticum* and *T. karavaievi* from Turkmenistan. We also observed the hibernation of the first and second instar larvae in *Lepisiota semenovi* from the same place.

Diapause in early instars is also typical for many *Camponotus* species. Larvae of European *C. vagus* always hibernate in the first instar [107]. Mintzer observed the same in incipient colonies of seven American *Camponotus* species (*C. clariothorax, C. festinatus, C. laevigatus, C. modoc, C. planatus, C. rasilis* and *C. vicinus*) under the laboratory conditions [119].

#### *6.2.2. Diapause in middle (2nd–4th of 5–6) instars (Camponotus s. str.)*

Sanders indicated that in *C. herculeanus, C. noveboraceus* and *C. pennsylvanicus,* living inside tree trunks in southern Canada, hibernating larvae could be divided into two distinct size classes, i.e. they were clearly at different instars, unlike the *Camponotus* species listed above [104, 105]. We observed the hibernation of larvae of the second, third and in a small number of the fourth instar in our experiments on *C. herculeanus, C. japonicus* and *C. ligniperda,* all belonging to the same subgenus *Camponotus s. str*. The first instar larvae never overwintered. In the natural nests of *C. herculeanus* in early spring, we also found mainly larvae of the second and third instar, but did not find larvae of the first and fifth instars at all.

#### *6.2.3. Diapause in the late (3rd–4th of 4) instars (Harpagoxenus, Leptothorax, Temnothorax and Messor)*

According to the literature data, the larvae of *Messor incorruptus* hibernate in the last two instars [114]. Delage noted that among the overwintering brood of *M. capitatus* there were no larvae of the first instar [106]. In the laboratory, we observed the hibernation of *M. denticulatus, M. intermedius, M. subgracilinodis* and *M. structor* and found that their larvae overwintered in the last (third and fourth) instars. In the nests of *M. denticulatus* and *M. structor* from Turkmenistan in the early spring, we always found only larvae of the last two instars.

Our investigations demonstrated that in *Leptothorax* and *Harpagoxenus sublaevis* larvae hibernated in the third and fourth instars, and larvae of the latter instar were clearly predominant among them. Only very rarely single larvae of the second instar could be found in overwintering colonies of these species. Larvae of *Temnothorax* may hibernate at all three instars, but larvae of the last instar obviously predominate among overwintering ones, for example, in *T. lichtensteini, T. nylanderi, T. parvulus* [69, 112], *T. grouvellei* [92], *T. melas* [75] and *T. monjanzei* [74]. We observed the same in *T. unifasciatus* and *T. tuberum.*

#### *6.2.4. Diapause in the last (usually 3rd) instar (Manica, Diplorhoptrum, Leptanilla, Monomorium, Myrmica, Tetramorium)*

Only larvae of the third (last) instar hibernate in all *Myrmica* species studied [109, 120–122]. Our studies on *M. bessarabica, M. kamtschatica, M. lacustris, M. rubra, M. ruginodis, M. sabuleti, M. scabrinodis* and *M. transsibirica* confirmed this. Larvae overwinter in the latter instar also in *Leptanilla japonica,* which inhabit the subtropical regions of Japan [73]. Our research work allowed us to supplement this group with the following species: *Monomorium gracillimum, M. ruzskyi, Manica rubida, Solenopsis celatum, S. fugax* and all *Tetramorium* species.

Larvae of the ants, hibernating at the last instar, are for the most part far from the completion of development, i.e. they are at the beginning or in the middle of this stage. To achieve the size required for pupation, they usually need a fairly long period of growth after overwintering. This fully applies to the largest of overwintering larvae that develop in spring into alate reproductives. An exception to this rule *is Leptothorax acervorum* and probably other species of the same subgenus, in which many of the larvae of reproductives almost complete their development before the winter.

*6.2.5. Diapause in all (of 3–6) instars (sometimes except the first instar) (Aphaenogaster, Crematogaster, Lasius, Paratrechina, Camponotus, Tanaemyrmex)*

This group includes *Lasius flavus* [84], *L. alienus* and *L. niger* (own data), *Paratrechina flavipes* [110], *Crematogaster bogojawlenskii* (own data) and some species of *Camponotus*. An exact study of the instar composition of overwintering larvae of *C. aethiops* in France showed that 85% of them were larvae of the first and second instars, 5% of the third to fourth and 10–15% of the fifth instar [108]. The results of our experiments and spring excavation of nests in nature confirmed that the overwintering brood of this species was represented by larvae of all instars except the first, with the obvious dominance of the second instar. We have shown that larvae of *C. bactrianus* and *C. xerxes*, species of the subgenus *Tanaemyrmex* as well, also hibernated in all instars except the first. Perhaps this feature is typical for the entire subgenus.

### 6.3. Seasonal cycles of the *Formica* type

According to the literature data and own observations, this annual cycle is typical for the ants of the entire tribe Formicini (genera *Alloformica, Cataglyphis, Formica, Proformica, Rossomyrmex*). This fact was firstly determined by Holmquist on *Formica ulkei* [123, 124] and later repeatedly confirmed by other researchers for same [125] and other *Formica* species: *F. fusca* [7], *F. haemorrhoidalis* [126], *F. japonica* [127] and *F. yessensis* [128]. Annual developmental cycles of ants of the subgenus *Formica s. str.* have been studied in detail by many authors (for example, see [129–132]). There are also detailed experimental data for *Cataglyphis cursor* [133–135]. Many species of genera *Alloformica, Cataglyphis, Formica* and *Proformica* have been studied in Central Asia by Dlusskii [63, 136]. The seasonal cycle of the slave-maker ant *Rossomyrmex proformicarum* in the deserts of Semirechie (Republic of Kazakhstan) was described by Marikovskii [137]. Our experiments and field observations made it possible to add to this list *Cataglyphis aenescens, C. emeryi, C. nodus, C. pallida, Formica cinerea, F. clara, F. cunicularia, F. gagatoides, F. lemani, F. picea, F pratensis* and *Proformica epinotalis.*

Outside the tribe Formicini, annual cycles without wintering brood were found in *Dolichoderus plagiatus, D. pustulatus* [138] and *D. quadripunctatus* [139, 140] and in American harvester ants *Pogonomyrmex occidentalis* and *P. montanus* [59, 141, 142]. *Prenolepis impairs* from the north of Florida also overwinters without brood but has completely special annual cycle [60, 62]. These ants are active outside the nests and foraging from November to March, i.e. during the winter months; the rest of a year the nests are closed. During the foraging period, workers inside the nest accumulate reserve nutrients in the fat body and gradually become physogastric. In spring, their mass exceeds the normal by 2–3 times. During spring and summer, the ants stay in the nests in inactive state. In September, the queens lay a large number of eggs, and then their ovaries again become dysfunctional. The ants feed the larvae emerging from eggs only at the expense of fat stocks of physogastric workers. Both workers and winged reproductives grow up from this single batch of brood. By the time of the resumption of foraging, there is no brood in the nests.

We found the annual cycle of *Formica* type in *Ponera coarctata* from the southern coast of the Crimea (Karadag). Investigation of several dozen nests in early September demonstrated that brood, except for a small number of pupae, was no longer present in them. In the colonies

http://dx.doi.org/10.5772/intechopen.75388

transferred from natural nests to the laboratory, the eggs did not appear even when the ants were kept at a long day photoperiod and at temperatures of 25–28°C for 2 months. Thus, the reproductive diapause of queens in this species was as stable as that of *Formica* ants. According to Talbot, *Ponera pennsylvanica* in Missouri (subtropics) also hibernated without brood [58]. Similar seasonal cycle was described for another Ponerine ant, *Odontomachus brunneus* from the north of Florida (subtropics) [143]. This species spend half of a year (from November up to April) without any brood in the nests. This allows us to assume that queens of *Odontomachus brunneus* have winter reproductive diapause.

## 7. Two types of the regulation of heterodynamic annual cycles

The main characteristic feature of homodynamic and quasi-heterodynamic types of annual cycles is a purely exogenous control of the development, and the key factor is environmental temperature. Heterodynamic species have a much more complex regulation of seasonal development, usually based on a combination of exogenous and endogenous mechanisms. We divided all heterodynamic ants into two groups, which differ substantially in the principles of the regulation of the annual cycle, exogenous-heterodynamic species and endogenous-heterodynamic species [18].

The first of them is characterized by the possibility of continuous and unlimited development under optimal conditions. The diapause is optional and occurs only when the temperature is lowered. Such annual cycles we call exogenous-heterodynamic. They are typical for all species of the genera *Messor, Monomorium, Solenopsis* and *Tetramorium* we have studied and also for *Camponotus xerxes* and *Tapinoma karavaievi*. At temperatures above 25°C, as well as at diurnal thermal periods 20/30°C, they all behave like true homodynamic species. Under these conditions, we observed continuous development without any signs of deterioration in experimental colonies for two or more years [144]. At the same time, at temperatures below 23–25°C, the egg laying and the larval pupation soon stop, or only pupation (in some species of *Messor*, which can hibernate with eggs), i.e. development finishes under the influence of suboptimal temperatures. If the temperature is then raised, the development will resume after a while. However, it can again be blocked by lowering the temperature, and once again stimulated by its increase. Similar experiments can be repeated with the same colony of ants many times and oft with the same result [144].

Thus, exogenous-heterodynamic species are distinguished by facultative winter diapause in larvae and queens. Developmental delays in a colony are purely exogenous and ensue as a straight reaction to the influence of external environmental factors, primarily, of temperature when it becomes not optimal for the development. Moreover, inhibition of development is unstable and easily disrupted when the temperature rises. However, these developmental delays are not just a consequence of cold coma but, namely, are the form of diapause (for more details, see [18]). This is not a state of elementary quiescence as in species with quasi-heterodynamic annual cycles, because this kind of diapause starts when environmental temperature still significantly exceeds the developmental threshold. Other essential feature is that the diapause arises not directly after the temperature decline but with some lag. Additionally, this diapause is invertible and may be terminated or induced again several times by consistent rising or decreasing the temperature, but each time after a little delay. The second important property of exogenous-heterodynamic species is the distinct change of their reaction to temperature during overwintering as a result of cold reactivation. After natural overwintering or after exposure in a refrigerator to 3–5°C during 2–3 months, the development and pupation of larvae recommenced and proceeded for a long period at 20°C and even at 18°C (in some *Tetramorium*) in all species studied [144]. This difference in the reaction to temperature is another indication of the existence of diapause in exogenous-heterodynamic species.

Most of the ants from temperate zone belong to the second group of species which is characterized by endogenous-heterodynamic annual cycles. The diapause arises due to internal factors (endogenous timer) and no external conditions can prevent it [18, 145]. Even under long day conditions and optimal temperatures, including the diurnal thermal periods, which are the most favorable thermal conditions for ants [146–148], the development in colonies of these species necessarily ceases, and the phase of dormancy in the annual cycle begins.

Thus, the diapause of endogenous-heterodynamic species is obligatory for the colony which has an internally limited intrinsic seasonal cycle of brood rearing. External environmental factors, such as temperature and photoperiod, also participate in the regulation of annual cycles of these species, playing a corrective role, i.e. in varying degrees adjusting the duration of the cycle to the climatic peculiarities of a particular summer season. But the regulation of the cycle is still based on processes that are endogenous for the colony [18, 145]. According to our data, the following species living in temperate climate belong to the group with endogenous-heterodynamic annual cycles: *Aphaenogaster, Camponotus, Cataglyphis, Crematogaster, Formica, Harpagoxenus, Lasius, Lepisiota, Leptothorax, Manica, Myrmica, Plagiolepis, Ponera, Proformica* and *Tapinoma* [18, 80, 96, 145–147, 149–151]. Analysis of the literature data from our conceptual positions allows us to classify the following species as endogenous-heterodynamic: *Aphaenogaster subterranea* [152], *Camponotus herculeanus, C. ligniperda, C. noveboraceus, C. pennsylvanicus, C. vagus* [105, 107, 153], *Cataglyphis cursor* [133–135], *Crematogaster scutellaris* [154–157], *Formica sanguinea* [158], *F. ulkei* [159], the genera *Leptothorax* and *Temnothorax* (with the exception of subtropical ones) [69, 75, 85, 91, 112], *Myrmica rubra* [82, 160–162], *Plagiolepis pygmaea* [103] and *Odontomachus brunneus* [143].

The gradual decrease of a colony capability to produce new eggs and to grow up non-diapausing larvae takes place during the summer season. At the same time, there is the increase of the bias for diapause as a consequence of the ongoing endogenous physiological and social processes. Moreover, the ant colony gradually acquires the sensitivity to the day length (to the photoperiod). The increasing photoperiodic sensitivity of a colony strongly changes the reaction to temperature. Because of these processes, at the end of the summer, the decrease of temperatures and the shortening of the day length (in some species only) contribute to the onset of diapause, thereby reducing the period of egg laying and larval development. The same impact of external factors to the colony's life cycle was found in all ant species that we studied [80, 96, 146, 163–165]. So, the duration of a colony's annual cycle of brood rearing in nature is controlled both by an endogenous timer and by exogenous environmental cues, such as temperature and photoperiod (in some species). These environmental conditions adjust the date of diapause onset to the climatic features of a given year.

Temperature control of diapause is the most universal in ants. The higher temperatures delay the onset of diapause both in larvae and adults, whereas the lower temperatures always accelerate the process of the beginning both in larvae and adults. On the contrary, photoperiodic control of diapause is unexpectedly uncommon among ants. For the first time, the photoperiodic responses were revealed in *M. rubra* and *M. ruginodis* [163]. It has been shown that the diapause in larvae and queen appeared more quickly when the day length decreased in the range from 16 to 13 h [163, 164]. The diapause both in queens and larvae ceased when inactive ant colonies in autumn condition were exposed to day lengths of 15 h [166]. Later on, Hand [167] and Brian [168] confirmed the presence of the reaction to photoperiodic conditions in *M. rubra*.

The subsequent extensive investigations nonetheless have shown that apart from *Myrmica* only few ant species used photoperiod as an ecological cue which controls all life processes in a colony: development, oviposition and diapause onset. We have found that photoperiodic conditions affected the induction of diapause only in *Aphaenogaster sinensis* [80] and *Lepisiota semenovi* [169]. Additionally, we have observed higher incidence of diapause in larvae under short-day conditions than in larvae under long-day conditions in *Camponotus herculeanus*, *Leptothorax acervorum* and *Manica rubida*. Thus, the genus *Myrmica* represents a rather curious exception among temperate ants since all *Myrmica* species studied so far possess clear-cut photoperiodic responses, which control the induction and termination of diapause [149, 150, 163, 164, 170, 171].

Thereby, the seasonal development of most ant species from temperate climate regions depends on inner timer in combination with environmental temperature, which triggers the onset of diapause. The environmental factors can modify the duration of the seasonal brood-rearing cycle within broad range in most species of the genera *Aphaenogaster*, *Crematogaster*, *Lasius*, *Myrmica* and *Tapinoma*. For example, suboptimal temperatures of 17–20°C and short days vastly bring closer the beginning of diapause in larvae and queens of *Myrmica rubra* and *M. ruginodis* [163]. Whereas at long days and a temperature of 25°C, egg laying and the development and pupation of rapid brood larvae continue for several months without interruption [149, 164].

The seasonal cycle of oviposition and development in other species is controlled predominantly by the endogenous mechanisms. In these ants, the moment of diapause onset depends only slightly on environmental conditions. Temperature hardly modifies the intrinsic length of the queens' oviposition period in all studied *Formica* species [96, 165]. The annual brood-rearing cycle in *Cataglyphis* species and the species from subgenera *Camponotus* s. str. and *Leptothorax* s. str. is also relatively independent of the environment.

## 8. The regulation of diapause in larvae

The diapause of larvae in ants is facultative in most case, i.e. a given larva can either develop directly or fall into a diapause depending on the circumstances. Temperature can affect larval development and induce diapause both directly and through the nurse workers. Detailed studies performed on *M. rubra* have made it possible to prove that ant larvae themselves were not able to sensate the photoperiodic conditions in which they were placed [172, 173].

Actually, the development of larvae was under the influence of workers whose physiological state and behavior strongly depended on the day length because they received the information about the photoperiodic conditions from the external environment. In laboratory experiments, we easily obtained the workers and larvae in alternative physiological states either by keeping the ant colonies under long-day conditions from the spring and by activating them with long-day photoperiods in autumn (getting of physiologically active, non-diapause individuals) or by exposing the ant colonies to the influence of short-day conditions during 1 month (getting of physiologically inactive, diapausing individuals). We have shown that non-diapause workers induced growth and pupation of summer larvae and interrupted the diapause of autumn larvae, while diapausing workers were incapable to support a high growth rate of larvae and forced them to fall into a diapause [13, 170]. We examined the characteristic features of social control by workers over larval development in several ant species and created experiments using workers and larvae in alternative physiological states. The experimental scheme included four sets: (1) spring state: physiologically active workers (after hibernation) with physiologically active larvae; (2) termination of larval diapause: physiologically active workers with diapausing larvae; (3) induction of larval diapause: diapausing workers with physiologically active larvae and (4) autumn state: diapausing workers with diapausing larvae.

The results of these experiments demonstrated that ant species studied fundamentally distinguished from each other [13–16, 170, 174]. In *Camponotus herculeanus, C. japonicus* and several *Tetramorium* species not overwintered physiologically inactive diapausing larvae, which were provided with food by spring physiologically active workers, developed rapidly and pupated within a short period, whereas overwintered larvae placed into the nests with autumn diapausing workers did not develop and pupate at all or only a few of them pupated sometimes. Thus, the workers of these ants exercise full control over the development and the diapause of their larvae. However, in *Leptothorax acervorum*, we found an entirely opposite situation: autumn workers could not prevent development of spring larvae and they all pupated. At the same time, overwintered workers stimulated the development of less than half of diapausing larvae. In *Myrmica rubra, M. ruginodis, M. lobicornis* and *Lasius niger*, we observed an intermediate situation: only some spring larvae pupated when fed by autumn workers, and also far from all autumn larvae finished the development under the care of spring workers.

Thus, the forms of social influence of workers on larval diapause are diverse in ants and range from nearly absolute control (in *Camponotus* and *Tetramorium*), when the physiological state of workers completely defines the fate of larvae, to rather weak effects when diapausing workers are unable to prevent the pupation of most of overwintered larvae, and spring workers are capable of inducing the development and pupation of only a few diapausing autumn larvae (in *Leptothorax*). In most species (*Lasius, Myrmica*), however, the intermediate variants of diapause control are realized. There is some evidence that *Myrmica* workers can manipulate the development of larvae via changing the intensity of tactile stimulation and the frequency of feedings [12, 13, 170, 174]. Probably, in the cases when larvae hibernate at younger instars (as in *Camponotus* s. str.) and need a long period of growth to complete development, diapausing workers cannot provide them with the necessary food. At the same time, spring workers are able to effectively stimulate the development of young diapausing larvae. On the other hand, large larvae, overwintered in the last instar (as in *Leptothorax*), can finish the development even without receiving enough food from diapausing workers. However, the diapause of these larvae is deeper and more durable, and the spring workers cannot break it.

In many ant species with endogenous- and exogenous-heterodynamic seasonal cycles, the larvae fall into a diapause in the last larval instar. These diapausing larvae continue to feed and to grow slowly and can attain a significantly larger size before overwintering. This larval growth in diapause state is very important for the process of caste differentiation in *Myrmica* [86]. Continuing to increase in the size the diapausing larvae in reality do not develop progressively, as far as the differentiation and enlargement of their imaginal buds do not occur. Only these well-grown large diapausing larvae potentially may become the female reproductives in spring [86]. Subsequently, Plateaux [69] described the identical phenomenon for *L. nylanderi*. In accordance with the results of our studies, slow growth of larvae in diapause state is an attribute of the ant species from the genera *Camponotus* (some species), *Crematogaster, Lasius, Lepisiota, Leptothorax, Manica, Messor, Monomorium, Myrmica, Solenopsis* and *Tetramorium*. The physiological nature and types of diapause in ants are examined in the paper of Kipyatkov [18].

## 9. Evolution of annual cycles of development in ants

Problems of the origin and evolution of the diapause and seasonal cycles of insect development attracted the attention of many researchers (for example, see [175–177]). The main conclusion reached by most authors is that the evolution of seasonal adaptations occurs largely beyond direct connection with the phylogeny of taxa, and all the elements and parameters of seasonal development known to us, including the diapause itself, the types of cycles, photoperiodic reactions and other regulatory mechanisms, repeatedly, independently and in a variety of ways arise in the evolution of insects. One and the same goal of adapting to certain seasonal environmental conditions can be reached in a variety of ways by combining a whole range of known (or even not yet described) physiological mechanisms [176, 177]. Such a clearly expressed diversity of evolutionary solutions makes it very difficult to analyze the possible ways of the evolution of seasonal development cycles, even within not very large groups of organisms. Nevertheless, some authors proposed various specific sequences of evolutionary events to explain the origin and development of photoperiodic reactions and other physiological mechanisms controlling the diapause and other seasonal adaptations [175]. The most promising approach to the problem of the evolution of seasonal adaptations can be the identification of basic adaptation syndromes, for example, structural types of the annual cycles, diapause forms, ways of synchronizing the cycle with the seasonal rhythm of the climate, etc. Then, it may be productive to look for correlations between all these adaptations and possible ties with the specific environmental conditions in which they are realized and also to analyze the occurrence of the phenomena under study within different taxa.

More and more data appear on the role of diapause in the regulation of the life cycles of insects in tropics. In these regions, the diapause does not always prove to be adapted specifically to the extreme conditions of the abiotic environment, but is often associated with seasonal variations in food availability, migration and reproduction processes. This makes it possible to speak with confidence about the tropical origin of the diapause in many insects [70, 176]. Moreover, the winter hibernation of insects is a relatively recent evolutionary acquisition that arose only after the formation of a glacial climate on the Earth [44]. The available facts do not completely exclude the hypothesis of the possible occurrence of the diapause by the gradual deepening of the developmental delays that first arise exogenously under the direct influence of cold, dryness, lack of food or other unfavorable conditions. The diapause is a fairly simple adaptation from the point of view of the possibility of forming its regulatory hormonal mechanisms, and therefore it could arise in evolution many times and in different ways [176, 177].

Turning to the analysis of the main trends in the evolution of the annual developmental cycles in ants, two most important and closely related questions should be pointed out. First, it is the origin of different forms of diapause, and secondly, possible ways of forming of endogenous-heterodynamic developmental cycles with obligate diapause. It should be borne in mind that the family Formicidae, in its evolutionary origin, is undoubtedly associated with the tropical regions of the Earth, where most of the species of ants live now. From tropical regions, these insects penetrated into zones with temperate climate, forming new species and taxa of higher rank [2, 63]. Another interesting issue, namely the evolution of the structure of the seasonal cycles in ants during their distribution to the north, has been discussed in detail in a special article of V. E. Kipyatkov [88].

It is possible to imagine at least two possible ways of the origin of heterodynamic annual cycles in ants living in temperate and cold climatic zones.

### 9.1. Subtropical (quasi-heterodynamic) path of evolution

We suppose the possibility of direct adaptation of homodynamic species which penetrate from the tropical regions into subtropics and further into the zone with temperate climate and with cold enough winter. In this case, they do not form a real diapause, and at first acquire the ability to overwinter in the state of a quiescent (cold coma) but suffer from more or less strong mortality, i.e. quasi-heterodynamic seasonal cycles appears. The diapause evolves later and seasonal development becomes exogenously heterodynamic.

The reality of this path of evolution is almost unquestionable. It has been illustrated above by a number of examples of quasi-heterodynamic seasonal cycles, in particular, by the example of two *Pheidole* species studied experimentally. Some of the quasi-heterodynamic species (*Linepithema humile, Solenopsis invicta, S. richteri*) are known to have penetrated recently into areas with subtropical and temperate climate. They represent a magnificent model of the evolution of heterodynamic cycles, but practically were not investigated experimentally. It is completely unclear; for example, do the populations of *S. invicta* that penetrated far enough to the north in the USA, shape some kind of a diapause, or in winter the development delays because of a quiescent, like in the populations from subtropics? The same question applies to the populations of *L. humile* in Southern France. An experimental study of these ants, as model systems, would make it possible to test the possibility of quite quickly (over several decades) occurrence of a diapause in quasi-heterodynamic species.

Probably, quasi-heterodynamic annual cycles with the brood death in late autumn arise in the evolution of many homodynamic ants, which penetrate from tropics into subtropics and further into regions with a warm temperate climate. The nomadic ants *Neivamyrmex* on the north of its distribution area in the USA live in regions with a fairly cold winter. Investigations were conducted on *N. nigrescens* by Schneirela in the southeast of Arizona at altitudes of about 1660 m above sea level [31]. He has shown that in the autumn, when the nights became cold, these ants with nocturnal activity ceased foraging activity. The lack of food compelled the queen to stop egg laying, the workers destroyed the remaining brood and adult ants hibernated in a shelter. In the spring when the temperature rose, the queen started oviposition. When larvae appeared, the ant colony gradually restored the cycle of nomadic behavior and brood rearing typical for the summer period [31]. Schneirela assumed the presence of direct influence of temperature on the development of these ants. However, this idea was not confirmed by experiments.

The second real way of verification of our assumption is to study the species taxonomically close to exogenous-heterodynamic ants from the zone with warm temperate climate, but inhabiting subtropics and tropics. The most promising in this respect are the genera *Monomorium* and especially *Tetramorium*. All of the species of these genera, which we studied, inhabit the temperate climate zone and are exogenous-heterodynamic.

If the onset of a diapause and exogenous-heterodynamic development on the basis of quasi-heterodynamic seasonal cycles seems quite plausible, then the possibility of further evolution in this direction toward the formation of endogenous-heterodynamic cycles with obligate diapause is far from obvious. We do not have at present any definitive evidence of the reality of such an evolution, but in our experiments we found distinct manifestations of endogenous regulation of development in most exogenous-heterodynamic species [144]. The possibilities of evolutionary transition from exogenous- to endogenous-heterodynamic development within the group of closely related species are indirectly confirmed by the following three examples.

Two closely related species of the genus *Tapinoma* from Turkmenistan were investigated in sufficient detail experimentally and under natural conditions. *T. karavaievi* is a species widespread in the plains of Central Asia. It does not rise to the mountains higher than 600 m above sea level, i.e. lives in a fairly warm temperate climate [136]. Almost in all parameters of the regulation of seasonal development, this species is exogenous-heterodynamic. In the most optimal temperature conditions, we observed almost non-stop development in its colonies. The diapause of larvae in *T. karavaiev*, as a rule, terminates when the temperature rises [144]. Nevertheless, the endogenous regulatory component was distinctly expressed in this species at lower temperatures, and under conditions of free temperature selection in the temperature gradient installation, we observed a smoothed spontaneous rhythm of development [145]. Another species studied, *T. erraticum*, is common in southern Europe and the Caucasus, and in Turkmenistan occurs only in the mountains, i.e. clearly gravitates toward a cooler climate than *T. karavaievi* [136]. Experiments have shown that *T. erraticum* belonged to the group of endogenous-heterodynamic species and had a very stable winter larval diapause, which could not be disturbed by a simple increase in temperature.

The second example is a pair of taxonomically related species of the genus *Monomorium*, *M. kusnezovi* and *M. ruzskyi*. Due to the existence of significant geographical variability in morphology for a long time, many researchers have combined these ants into one species under different names and confused them [136]. The gradual accumulation of collection materials made it possible to find out that everywhere on the plains of Central Asia there is one variegated in color species, *M. kusnezovi*. Its mountain populations differ from the desert ones not only in smaller sizes and dark colors, but also in some peculiarities of biology [63]. Later, large differences between the queens of these two forms were found and the mountain form was described as *M. ruzskyi* [136].

Our observations have shown that *M. kusnezovi* and *M. ruzskyi* often met together in the foothills and lowlands of the Kopetdag. However, despite the taxonomic proximity of these two species, their annual cycles differ significantly. *M. kusnezovi* is quasi-heterodynamic, i.e. the diapause of the brood and the reproductive diapause of queens are absent, so eggs and larvae of all ages remain in the nest during the winter and a significant part of them perish. At the same time, the annual cycle of *M. ruzskyi* is of *Myrmica* type, i.e. it is endogenous-heterodynamic and is characterized by a fairly stable diapause in larvae of the last (third) instar and in queens. Therefore, in the colonies of this species, only the larvae of the last instar hibernate, and they are always numerous. Thus, the features of the structure and regulation of the annual cycle make it possible to differentiate these closely related species no worse than the morphology of their queens.

We studied two *Camponotus* species, *C. xerxes* and *C. aethiops*, which belonged to the subgenus *Tanaemyrmex*, but differed dramatically in regulation of seasonal development. *C. aethiops* possesses an obligate, very stable diapause and endogenous-heterodynamic regulation of the annual cycle, while in *C. xerxes* there is an unstable diapause and mainly exogenous regulation. At a temperature of 25–27°C, we observed continuous development in the colonies of *C. xerxes* for more than 2 years. These differences are obviously due to the fact that *C. aethiops* lives in a colder climate (Central and Southern Europe, Crimea, Caucasus and mountains of Central Asia) than *C. xerxes* (the desert plains of Afghanistan, Iran and Turkmenistan).

### 9.2. Tropical (preadaptational) path of evolution

Extremely few examples of heterodynamic annual cycles in tropical ants allow us to assume that such species, already possessing diapause (preadaptation), could probably easily penetrate into subtropics and further into the temperate climate zone, using the ability to form a diapause for experiencing a cold winter. The causes for diapause emergence in tropical ants can be diverse. For example, it can be the necessity for survival during the arid or excessively wet seasons of a year, during the periods of shortage or inability to obtain food, etc. It is also possible that the ability to diapause arose as a way of solving internal problems for the colony related to the regulation of development, the processes of caste differentiation and reproduction.

http://dx.doi.org/10.5772/intechopen.75388

Moreover, in tropics, both exogenous and endogenous mechanisms of diapause regulation could be formed, which under the new conditions of a temperate climate, and could be used to synchronize the onset of a diapause with the beginning of a cold season of a year.

Unfortunately, this scheme is almost entirely speculative, first of all, because tropical species with heterodynamic annual cycles have not been investigated experimentally so far, and we know absolutely nothing about the prevalence and nature of diapause in tropical ants. The only argument indirectly confirming the possibility of the tropical origin of heterodynamic annual development in many ants is the rather wide prevalence of endogenous-heterodynamic cycles with obligate diapause among ants inhabiting subtropical and warm temperate climate zone. Among the 39 species of ants from Turkmenistan and the southern coast of the Crimea which we experimentally studied, 18 (46%) are endogenous-heterodynamic. An analysis of a few works, in which at least the simplest laboratory experiments were carried out, allows us to classify as endogenous-heterodynamic four more species from the zone with warm temperate and subtropical climate: *Plagiolepis pygmaea* [103], *Camponotus vagus* [107], *Temnothorax monjanzei* [74] and *T. melas* [75]. Are we correct to assert that in all these ants, the predominantly endogenous regulation of the annual cycle with the obligate diapause has evolved here in areas with a very warm or subtropical climate, or is it wiser to think that it originated in tropical ancestors of some of these species at least? The answer to this question is currently not possible due to lack of sufficient evidence.

### 9.3. Taxonomic position and seasonal adaptations

An example of close relationship between taxonomic position and seasonal adaptations is the Formicini tribe. All species studied use the strategy of concentrated brood rearing and are endogenous-heterodynamic with the obligate and very stable diapause of queens and workers and the apparent dominance of endogenous regulatory mechanisms. Probably, such a system of seasonal adaptations evolved already in the ancestors of this tribe.

All studied species of the genus *Myrmica* have a very similar structure of the seasonal development cycle [178, 179] and, according to our experiments, have photoperiodic regulation of a diapause rarely seen in other ants [18]. We have experimentally investigated the seasonal development of three species of the subgenus *Camponotus* s. str., *C. herculeanus, C. japonicus* and *C. ligniperda* and found almost complete similarity between them in all respects: endogenous-heterodynamic regulation, obligate and very stable diapause of queens and larvae, the same overwintering stages (second instar larvae) and, finally, absolute control by workers over the onset and cessation of larval diapause [16]. According to available literature data, *C. vagus* from the same subgenus has identical characteristics of the seasonal cycle [107]. Finally, all species of the genus *Tetramorium*, which we studied, are distinguished by exogenous-heterodynamic regulation of seasonal development and temperature-unstable diapause.

Other examples relate to fairly definite correlations between taxonomic position and overwintering stages in a number of ants genera. In all species of *Leptothorax* and *Messor*, larvae diapause and overwinter mainly in the last two instars; in *Myrmica* and *Tetramorium*, in the last instar, obligatory; in *Plagiolepis*, in the first three instars and in *Lasius*, in all instars.

Finally, all *Aphaenogaster* species studied so far do not have a diapause of queens and therefore hibernate with eggs and larvae in all instars. At the same time, directly opposite situations are known, when closely related ants have completely different seasonal adaptations. They are already mentioned above when comparing the following pairs of species: *T. karavaievi–T. erraticum, M. kusnezovi–M. ruzskyi* and *C. xerxes–C. aethiops.*

Thus, it can be argued that in the evolution of ants, correlations could arise between the nature of seasonal adaptations and the phylogeny of taxa, but no less common are the cases of the absence of such connections, i.e. the appearance of significant differences in the structure and regulation of the annual cycle between related species and, on the contrary, the parallel and independent formation of very similar adaptations in different phylogenetic branches.

## 10. Conclusions

1. Most tropical ants demonstrate homodynamic development. They do not exhibit any developmental delays and all-year round the ontogenetic stages from egg to pupa exist in their nests. Some of the quasi-heterodynamic species have permeated into the regions with warm temperate climate but a true diapause did not evolve. In these species, the brood development stops only at temperatures falling below the developmental threshold (consecutive dormancy). So, the ants spend the winter in the state of a quiescent (cold coma), while more or less high mortality rates are observed in their colonies. Most temperate and all boreal climate ants are true heterodynamic. They manifest a true deep winter diapause (prospective dormancy) in their annual cycle.

2. Heterodynamic ants use two main seasonal strategies with respect to brood rearing. The ants are more likely to follow the strategy of prolonged brood rearing. It is distinguished by the following features: (1) larval diapause is facultative and controlled by environmental (temperature, photoperiod) and social (worker care, queen influence, pheromones, etc.) factors; (2) only some larvae develop from egg to pupa within the same summer season without overwintering (this rapid brood, or summer brood, yields only workers); (3) a large proportion of larvae delay their development, continue to grow in autumn, overwinter in diapause and pupate the next summer (this slow brood, or winter brood, yields both workers and alates).

3. The strategy of concentrated brood rearing is distinguished by the following features: (1) larvae have no dormancy and complete their development during the summer; (2) the development of all brood stages is thus restricted to the growing season; (3) only queens and workers are able to undergo diapause and overwinter; (4) the colony thus passes the winter without brood. This strategy, however, is not the most common.

4. The forms of dormancy which were found in ants extend from elementary quiescence to deep diapause. In exogenous-heterodynamic species, the diapause is optional for larvae and queens. The diapause occurs as a result of a direct reaction to temperature decline in the autumn but at a moment when the temperatures still exceed the developmental threshold.

5. On the contrary in endogenous-heterodynamic ant species, the diapause is compulsory for the colony and occurs eventually under any conditions. Two main factors restrict and control the internal brood-rearing cycle in these species. They are the endogenous timer and environmental conditions, temperature and photoperiod (in some species). But environmental cues can only regulate in some degree the moment of the onset of diapause by accelerating or delaying this event. All cold climate ants have adult diapause, so that their queens and workers are capable for overwintering. Queens and some workers experience diapause several times in their life. On the contrary, the ability of larvae to undergo diapause is not universal in ants. This is a major factor in seasonal cycle evolution in these insects.
6. The diapause of larvae in ants is facultative in most case. Temperature can affect larval development and induce diapause both directly and through the nurse workers. The larvae appeared to be entirely insensitive to the direct influence of photoperiods. The forms of social impact on larval diapause by workers are diverse in ants and range from nearly absolute control when the physiological state of workers completely defines the fate of larvae, to rather weak effects when in experimental conditions diapausing workers are unable to prevent the development of most overwintered larvae, and spring workers are capable to induce pupation of only a few diapausing autumn larvae.
7. We can conclude that the similar seasonal adaptations could arise in ant evolution independently many times and usually are not tightly bound to the taxonomic position of species. Nevertheless, several examples of certain seasonal cycle traits clearly confined to specific ant taxa have been found.

## Acknowledgements

I am very grateful to students and members of the Department of Entomology, St. Petersburg State University, who participated in these long-term investigations in different years. This work was supported by grants from European Union INTAS program (94-2072), Russian Foundation of Basic Research (97-04-48987, 00-04-49003, 03-04-48854, 06-04-49383), Federal Program "Universities of Russia" (07.01.026, 07.01.327) and the Council for Grants from the President of the Russian Federation and for State Support of Leading Scientific Schools (00-15-97934, 2234.2003.4, 7130.2006.4).

## Author details

Elena B. Lopatina†

Address all correspondence to: elena.lopatina@gmail.com

Department of Entomology, Saint Petersburg State University, Russian Federation

† This review is dedicated to the memory of Professor V.E. Kipyatkov (1949–2012) who devoted his life to the study of seasonal cycles of development in ants.

## References


[1] Brian M. Social Insects. Ecology and Behavioural Biology. London and New York: Chapman and Hall; 1983. 377 p. DOI: 10.1007/978-94-009-5915-6

[2] Hölldobler B, Wilson E. The Ants. Cambridge, MA: Belknap Press; 1990. 732 p

[3] Yozhikov I. On the comparative ecology of social insects. Trudy Otdeleniya Ecologii Timiryazevskogo Nauchno-Issledovatel'skogo Instituta. 1929;**4(1)**:7-24 (in Russian)

[4] Headley A. Population studies of two species of ants, *Leptothorax longispinosus* Roger and *Leptothorax curvispinosus* Mayr. Annals of the Entomological Society of America. 1943;**36**:743-753. DOI: 10.1093/aesa/36.4.743

[5] Headley A. A population study of the ant *Aphaenogaster fulva ssp. aquia* Buckley (Hymenoptera, Formicidae). Annals of the Entomological Society of America. 1949;**42**:265-272. DOI: 10.1093/aesa/42.3.265

[6] Talbot M. Population studies of the ant *Myrmica schencki ssp. emeryana* Forel. Annals of the Entomological Society of America. 1945;**38**:365-372. DOI: 10.1093/aesa/38.3.365

[7] Eidmann H. Die überwinterung der Ameisen. Zeitschrift für Morphologie und Ökologie der Tiere. 1943;**39**:217-275. DOI: http://www.jstor.org/stable/43261901

[8] Brian M. The synchronisation of colony and climatic cycles. In: Proceedings of the 8th International Congress of the International Union for the Study of Social Insects, September 5-10, 1977, Wageningen. 1977. pp. 202-206

[9] Brian M. Regulation of sexual production in an ant society. In: Colloques Internationaux du Centre National de la Recherche Scientifique, Paris: 1967. No 173 Ed. 1968. pp. 61-74

[10] Brian M. Ants. London: Collins; 1977. 223 p

[11] Kipyatkov V. Distantly perceiving primer pheromone controls the diapause termination in the ant, *Myrmica rubra* L. (Hymenoptera, Formicidae). Journal of Evolutionary Biochemistry and Physiology 2001;**37**:405-416. DOI: 10.1023/A:1012926929059

[12] Kipyatkov V, Lopatina E. Behavior of worker ants during the feeding of larvae in *Myrmica rubra* (Hymenoptera, Formicidae). Zoologicheskii Zhurnal. 1989;**68**:50-59 (in Russian with English summary)

[13] Kipyatkov V, Lopatina E. Quantitative study of the behavior of the ant *Myrmica rubra* L. (Hymenoptera, Formicidae) relative to photoperiodic regulation of larval development. Entomologicheskoe Obozrenie. 1989;**68**:251-261 (in Russian with English summary; English translation in: Entomological Review. 1990;**69**:31-41)

[14] Kipyatkov V, Lopatina E, Pinegin A. Influence of workers and the queen on the onset and termination of the larval diapause in the ant *Lasius niger* (L.) (Hymenoptera, Formicidae). Entomologicheskoe Obozrenie. 1996;**75**:507-515 (in Russian with English summary; English translation in: Entomological Review. 1996;**76**:514-520)

[15] Kipyatkov V, Lopatina E, Pinegin A. Social regulation of development and diapause in the ant *Leptothorax acervorum* (Hymenoptera, Formicidae). Entomological Review. 1997;**77**:248-255

[16] Lopatina E, Kipyatkov V. Behavioural and physiological control by workers over the development of larvae in the seasonal life cycle of ant colony. In: Programme and Abstracts of the International Symposium "Life Cycles in Social Insects: Behavioural, Ecological and Evolutionary Approach" September 22-27, 2003; St. Petersburg, Russia: St. Petersburg University Press; 2003. pp. 40-42

[17] Bourke A, Franks N. Social evolution in ants. In: Krebs J, Clutton-Brock T, series editor. Monographs in Behavior and Ecology. Princeton, New Jersey: Princeton University Press; 1995. ix+529 p. DOI: 10.1046/j.1420-9101.1996.9061032.x

[18] Kipyatkov V. Seasonal life cycles and the forms of dormancy in ants (Hymenoptera, Formicoidea). Acta Societatis Zoologicae Bohemicae. 2001;**65**:211-238

[19] Roubaud E. Étude sur le sommeil d'hiver pre-imaginal des muscides. Bulletin Biologique De La France Et De La Belgique. 1922;**56**:455-544

[20] Roubaud E. Sommeil d'hiver cédant à l'hiver chez les larves et nymphes de muscides. Comptes Rendus de l'Académie des Sciences. 1922;**174**:964-966

[21] Roubaud E. La réactivation climatique et la distribution géographique des espèces. Compte Rendu de la Association Française pour l'Avancement des Sciences. 1925;**48**: 982-986

[22] Ackonor J. The nest and nesting habits of the ant *Cataulacus guineensis* F. Smith (Hymenoptera, Formicidae) in a Ghanaian cocoa farm. Insect Science and its Application. 1983;**4**:267-283. DOI: 10.1017/S174275840000126

[23] Basalingappa S, Kareddy V, Matapathi S, Gandhi M. Population studies on the ant, *Camponotus sericeus* Fabricius. Comparative Physiology and Ecology. 1986;**11**:32-37

[24] Basalingappa S, Kareddy V, Matapathi S, Gandhi M. Colony-founding and life-cycle in the ant *Camponotus sericeus* Fabricius (Hymenoptera: Formicidae). Uttar Pradesh Journal of Zoology. 1989;**9**:73-83

[25] Braun U, Hölldobler B, Peeters C. Colony life cycle and sex ratio of the ant Paltothyreus *tarsatus* in Ivory Coast. In: Lenoir A, Arnold G, Lepage M, editors. Les Insectes Sociaux. 12th Congress of the International Union for the Study of Social Insects, Paris, Sorbonne, August 21-27, Sorbonne, Paris 1994. Paris: Université Paris Nord; 1994. p. 330

[26] Baker G. The seasonal life cycle of *Anoplolepis longipes* (Jerdon) (Hymenoptera: Formicidae) in a cacao plantation and under brushed rain forest in the northern district of Papua New Guinea. Insectes Sociaux. 1976;**23**:253-261. DOI: 10.1007/BF02283900

[27] Haines I, Haines J. Colony structure, seasonality and food requirements of the crazy ant, *Anoplolepis longipes* (Jerd.), in the Seychelles. Ecological Entomology. 1978;**3**:109-118. DOI: 10.1111/j.1365-2311.1978.tb00909.x

[28] Curtis B. Observations on the natural history and behaviour of the dune ant, *Camponotus detritus* Emery, in the central Namib Desert. Madoqua. 1985;**14**:279-289

[29] Ward P. Ecology and life history of the *Rhytidoponera impressa* group (Hymenoptera: Formicidae). II. Colony origin, seasonal cycles, and reproduction. Psyche 1981;**88**:109-126. DOI: 10.1155/1981/39593

[30] Benois A. Incidence des facteurs écologiques sur le cycle annuel et l'activité saisonnière de la fourmi d'Argentine, *Iridomyrmex humilis* Mayr (Hymenoptera, Formicidae), dans la région d'Antibes. Insectes Sociaux. 1973;**20**:267-295. DOI: 10.1007/BF02223196

[31] Schneirla T. Army ants. In: Topoff HR, editor. A Study in Social Organization. San Francisco: W. H. Freeman and Company; 1971. XX+349 p

[32] Gotwald Jr W. Army ants. In: Hermann HR, editor. Social Insects Vol. 4. New York: Academic Press; 1982. pp. 157-254

[33] Rettenmeyer C, Chadab-Crepet R, Naumann M, Morales L. Comparative foraging by Neotropical army ants. In: Jaisson P, editor. Social Insects in the Tropics. Proceedings of the First International Symposium. Vol. 2. Paris: Presses de l'Université Paris XIII; 1983. pp. 59-73

[34] Terron G. Évolution des colonies de *Tetraponera anthracina* Santschi (Formicidae, Pseudomyrmecinae) avec reines. Bulletin Biologique de la France et de la Belgique. 1977;**111**:115-181

[35] Peacock A. Studies in Pharaoh's ant, *Monomorium pharaonis* (L.). 2. Methods of recording observations on artificial colonies. Entomologist's Monthly Magazine. 1950;**86**:129-135

[36] Peacock A. Studies in Pharaoh's ant, *Monomorium pharaonis* (L.). 4. Egg-production. Entomologist's Monthly Magazine. 1950;**86**:294-298

[37] Peacock A, Baxter A. Studies in Pharaoh's ant, *Monomorium pharaonis* (L.). 1. The rearing of artificial colonies. Entomologist's Monthly Magazine. 1949;**85**:256-260

[38] Peacock A, Baxter A. Studies in Pharaoh's ant, *Monomorium pharaonis* (L.). 3. Life history. Entomologist's Monthly Magazine. 1950;**86**:171-178

[39] Peacock A, Waterhouse F, Baxter A. Studies in Pharaoh's ant, *Monomorium pharaonis* (L.). 10. Viability in regard to temperature and humidity. Entomologist's Monthly Magazine. 1955;**91**:37-42

[40] Petersen Braun M. Untersuchungen zur sozialen Organisation der Pharaoameise *Monomorium pharaonis* (L.) (Hymenoptera, Formicidae). I. Der Brutzyklus und seine Steuerung durch populationseigene Faktoren. Insectes Sociaux. 1975;**22**:269-291. DOI: 10.1007/BF02223077

[41] Petersen-Braun M. Studies on the endogenous breeding cycle in *Monomorium pharaonis* L. (Formicidae). In: Proceedings of the 8th International Congress of the International

Union for the Study of Social Insects, September 5-10, 1977, Wageningen; 1977. pp. 211-212

[42] Kipyatkov V, Lopatina E. A Comparative study of thermal reaction norms for development in ants. Entomological Science. 2015;**18**:174-192. DOI: 10.1111/ens.12098

[43] Berndt K-P, Eichler W. Die Pharaoameise, *Monomorium pharaonis* (L.) (Hym., Myrmicidae). Mitteilungen aus dem Zoologischen Museum in Berlin. 1987;**63**:183-186. DOI: 10.1002/mmnz.19870630102

[44] Leather S, Walters K, Bale J. The Ecology of Insect Overwintering. Cambridge: University Press; 1993. X+255 p. DOI: 10.1017/CBO9780511525834

[45] Baldridge R, DeGraffenried J. *Pseudomyrmex* sp. (Hymenoptera: Formicidae) nesting in mimosa (*Albizia julibrissin* Dur.). The Southwestern Naturalist. 1988;**33**:112-114. DOI: 10.2307/3672102

[46] Rissing S. Annual cycles in worker size of the seed-harvester ant *Veromessor pergandei* (Hymenoptera: Formicidae). Behavioral Ecology and Sociobiology. 1987;**20**;117-124. DOI: http://www.jstor.org/stable/4599998

[47] Lennartz F. Modes of dispersal of *Solenopsis invicta* from Brazil into the continental United States, a study in spatial diffusion [thesis]. University of Florida; 1973. 242 p

[48] Lofgren C, Banks W, Glancey B. Biology and control of imported fire ants. Annual Review of Entomology. 1975;**20**:1-30. DOI: 10.1146/annurev.en.20.010175.000245

[49] Lofgren C. History of imported fire ants in the United States. In: Lofgren CS, Vander Meer RK, editors. Fire Ants and Leaf Cutting Ants: Biology and Management. Boulder, CO: Westview Press; 1986. pp. 36-47

[50] Shoemaker D, Ross K, Arnold M. Development of RAPD markers in two introduced fire ants, *Solenopsis invicta* and *S. richteri,* and their application to the study of a hybrid zone. Molecular Ecology. 1994;**3**:531-539. DOI: 10.1111/j.1365-294X.1994.tb00084.x

[51] Markin G, Dillier J. The seasonal life cycle of the imported fire ant, *Solenopsis saevissima richteri,* on the Gulf coast of Mississippi. Annals of the Entomological Society of America. 1971;**64**:562-565. DOI: 10.1093/aesa/64.3.537

[52] Horton P, Hays S. Occurrence of brood stages and adult castes in field colonies of the red imported fire ant in South Carolina. Environmental Entomology. 1974;**3**:656-658. DOI: 10.1093/ee/3.4.656

[53] Markin G, O'Neal J, Dillier J, Collins H. Regional variation in the seasonal activity of the imported fire ant, *Solenopsis saevissima richteri*. Environmental Entomology. 1974;**3**:446-452. DOI: 10.1093/ee/3.3.446

[54] Porter S. Impact of temperature on colony growth and developmental rates of the ant, *Solenopsis invicta*. Journal of Insect Physiology. 1988;**34**:1127-1133. DOI: 10.1016/0022-1910(88)90215-6

[55] Morrill W. Overwinter survival of the red imported fire ant in Central Georgia. Environmental Entomology. 1977;**6**:50-52. DOI: 10.1093/ee/6.1.50

[56] Markin G. The seasonal life cycle of the argentine ant, *Iridomyrmex humilis* (Hymenoptera: Formicidae), in South-California. Annals of the Entomological Society of America. 1970; **63**:1238-1242. DOI: 10.1093/aesa/63.5.1238

[57] Vargo E, Passera L. Gyne development in the argentine ant *Iridomyrmex humilis*: Role of overwintering and queen control. Physiological Entomology. 1992;**17**:193-201. DOI: 10.1111/j.1365-3032.1992.tb01199.x

[58] Talbot M. Population studies of ants in a Missouri woodland. Insectes Sociaux. 1957;**4**:375-384. DOI: 10.1007/BF02224157

[59] MacKay W. A comparison of the nest phenologies of three species of *Pogonomyrmex* harvester ants (Hymenoptera: Formicidae). Psyche. 1981;**88**:25-74. DOI: 10.1155/1981/78635

[60] Talbot M. Population studies of the ant, *Prenolepis imparis* say. Ecology. 1943;**24**:31-44. DOI: 10.2307/1929858

[61] Talbot M. Response of the ant *Prenolepis imparis* say to temperature and humidity changes. Ecology. 1943;**24**:345-352. DOI: 10.2307/1930536

[62] Tschinkel W. Seasonal life history and nest architecture of a winter-active ant, *Prenolepis impairs*. Insectes Sociaux. 1987;**34**:143-164. DOI: 10.1007/BF02224081

[63] Dlusskiy G. Murav'i Pustyn' [Ants of Deserts]. Moscow: Nauka; 1981. 230 p. (in Russian)

[64] Kupyanskaya A. Murav'i (Hymenoptera, Formicidae) Dal'nego Vostoka SSSR [Ants (Hymenoptera, Formicidae) of the USSR Far East]. Vladivostok: Far East Branch of Acad. Sci. USSR; 1990. 258 p. (in Russian)

[65] Passera L. Production des soldats dans les sociétés sortant d'hibernation chez la Fourmi *Pheidole pallidula* (Nyl.) (Formicidae Myrmicinae). Insectes Sociaux. 1977;**24**:131-146. DOI: 10.1007/BF02227167

[66] Passera L. La ponte d'oeufs préorientés chez la fourmi *Pheidole pallidula* (Nyl.) (Hymenoptera: Formicidae). Insectes Sociaux. 1980;**27**:79-95. DOI: 10.1007/BF02224522

[67] Passera L, Suzzoni JP. Le rôle de la reine de *Pheidole pallidula* (Nyl.) (Hymenoptera, Fromicidae) dans la sexualisation du couvain après traitement par l'hormone juvénile. Insectes Sociaux. 1979;**26**:343-353. DOI: 10.1007/BF02223553

[68] Murdock TC, Tschinkel WR. The life history and seasonal cycle of the ant, *Pheidole morrisi* Forel, as revealed by wax casting. Insectes Sociaux. 2015;**62**:265-280. DOI: 10.1007/s00040-015-0403-9

[69] Plateaux L. Sur le polymorphisme social de la Fourmi *Leptothorax nylanderi* (Förster). I. - Morphologie et biologie comparées des castes. Annales des Sciences Naturelles. (Zoologie). 1970;**12**:373-478

[70] Denlinger D. Dormancy in tropical insects. Annual Review of Entomology. 1986;**31**:239-264. DOI: 10.1146/annurev.en.31.010186.001323

[71] Ito F, Higashi S, Maeta Y. Growth and development of *Camponotus (Paramyrmamblys) kiusiuensis* Santschi colonies (Hymenoptera Formicidae). Insectes Sociaux. 1988;**35**:251-261. DOI: 10.1007/BF02224058

[72] Satoh T. Comparisons between two apparently distinct forms of *Camponotus nawai* Ito (Hymenoptera, Formicidae). Insectes Sociaux. 1989;**36**:277-292. DOI: 10.1007/BF02224881

[73] Masuko K. Behavior and ecology of the enigmatic ant *Leptanilla japonica* Baroni Urbani (Hymenoptera: Formicidae: Leptanillinae). Insectes Sociaux. 1990;**37**:31-57. DOI: 10.1007/BF02223813

[74] Cagniant H. Description de *Leptothorax monjanzei n. sp.* d'Algérie. Bulletin de la Société Entomologique de France. 1968;**73**:83-90

[75] Espadaler X, Plateaux L, Casevitz-Weulersse J. *Leptothorax melas*, n. sp., de Corse notes écologiques et biologiques [Hymenoptera, Formicidae]. Revue Française d'Entomologie. 1984;**6**:123-132

[76] Herbers J. The population biology of *Tapinoma minutum* (Hymenoptera: Formicidae) in Australia. Insectes Sociaux. 1991;**38**:195-204. DOI: 10.1007/BF01240969

[77] Chen Y, Tang J. Studies on colony structure and lifecycle of the spinned ant *Polyrhachis vicina* Roger. Zoological Research. 1989;**10**:57-63

[78] Chen Y, Tang J. Daily rhythm and seasonal activity of the weaver ant, *Polyrhachis vicina*. In: Proceedings of 19th International Congress of Entomology; 28 June–4 July 1992; Beijing, China. 1992. p. 241

[79] Laskis K, Tschinkel W. The seasonal natural history of the ant, *Dolichoderus mariae*, in northern Florida. Journal of Insect Science. 2009;**9**(2):26. DOI: 10.1673/031.009.0201

[80] Kipyatkov V, Lopatina E. Seasonal development of *Aphaenogaster sinensis* in the South Primorye: A new type of seasonal cycles in ants. Zoologicheskii Zhurnal. 1990;**69**:69-79 (in Russian with English summary, English translation in: Entomological Review. 1990;**69**:72-81)

[81] Brian M. Summer population changes in colonies of the ant *Myrmica*. Physiologia Comparata et Oecologia. 1951;**2**:248-262

[82] Brian M. Serial organization of brood in *Myrmica*. Insectes Sociaux. 1957;**4**:191-210. DOI: 10.1007/BF02222153

[83] Traniello J. Population structure and social organization in the primitive ant *Amblyopone pallipes* (Hymenoptera: Formicidae). Psyche. 1982;**89**:65-80. DOI: 10.1155/1982/79349

[84] Peakin G. The growth and development of the overwintering larvae in the ant *Lasius flavus* (Hymenoptera, Formicidae). Journal of Zoology. 1985;**205**:179-189. DOI: 10.1111/j.1469-7998.1985.tb03527.x

[85] Plateaux L. Sur le polymorphisme social de la Fourmi *Leptothorax nylanderi* (Förster). II. – Activité des ouvrières et déterminisme des castes. Annales des Sciences Naturelles (Zoologie). 1971;**13**:1-90

[86] Brian M. Studies of caste differentiation in *Myrmica rubra* L. 3. Larval dormancy, winter size and vernalisation. Insectes Sociaux. 1955;**2**:85-114. DOI: 10.1007/BF02224096

[87] Wilson E. The Insect Societies. Cambridge, MA: Belknap Press; 1971. 548 p

[88] Kipyatkov V. The evolution of seasonal cycles in cold-temperate and boreal ants: Patterns and constraints. In: Kipyatkov V, editor. Life Cycles in Social Insects: Behaviour, Ecology and Evolution. St. Petersburg: St. Petersburg University Press; 2006. pp. 63-84

[89] Martin P. Cycle annuel de *Leptothorax unifasciatus* (Latr.) elevé en laboratoire. Résultats preliminaries. Actes des Colloques Insectes Sociaux. 1988;**4**:169-175

[90] Suzzoni J-P, Grimal A, Passera L. Modalité et cinétique du développement larvaire chez la fourmi *Camponotus aethiops* Latr. Bulletin de la Société Zoologique de France. 1986;**111**:113-121

[91] Buschinger A. The role of daily temperature rhythms in brood development of ants of the tribe Leptothoracini (Hymenoptera; Formicidae). In: Wieser W, editor. Effects of Temperature on Ectothermic Organisms. Berlin: Springer Verlag; 1973. pp. 229-232. DOI: 10.1007/978-3-642-65703-0

[92] Espadaler X, Du Merle P, Plateaux L. Redescription de *Leptothorax grouvellei* Bondroit, 1918. Notes biologiques et écologiques (Hymenoptera, Formicidae). Insectes Sociaux. 1983;**30**:274-286. DOI: 10.1007/BF02223985

[93] Brian M. The growth and development of colonies of the ant *Myrmica*. Insectes Sociaux. 1957;**4**:177-190. DOI: 10.1007/BF02222152

[94] Danks H. Arctic Arthropods: A Review of Systematics and Ecology with Particular Reference to the North American Fauna. Ottawa: Entomological Society of Canada; 1981. vii+608 p

[95] Danks H. Life cycles in polar arthropods – Flexible or programmed? European Journal of Entomology. 1999;**96**:83-102

[96] Kipyatkov V, Lopatina E. The regulation of annual cycle of development in the ants of the subgenus *Serviformica* (Hymenoptera, Formicidae). In: Kipyatkov VE, Editor. Proceedings of the Colloquia on Social Insects. Vol. 2. Russian-speaking Section of the IUSSI, Socium: St. Petersburg; 1993. pp. 49-60

[97] Talbot M. Populations and hibernating conditions of the ant *Aphaenogaster (Attomyrma) rudis* Emery (Hymenoptera: Formicidae). Annals of the Entomological Society of America. 1951;**44**:302-307. DOI: 10.1093/aesa/44.3.302

[98] Mizutani A, Imamura S. Population and nest structure in the ant *Aphaenogaster japonica* Forel, in Sapporo, Japan. Kontyû. 1980;**48**:241-247

[99] Thomas M. Seasonality and colony-size effects on the life-history characteristics of *Rhytidoponera metallica* in temperate South-Eastern Australia. Australian Journal of Zoology. 2003;**51**:551-567. DOI: 10.1071/ZO03037

[100] Passera L, Dejean A. Étude de la ponte à basse température de *Temnothorax recedens* (Nyl.) et de quelques espèces du genre *Leptothorax* (Formicidae, Myrmicinae). Insectes Sociaux. 1974;**21**:407-415. DOI: 10.1007/BF02331568

[101] Talbot M. Population studies of the slave-making ant *Leptothorax duloticus* and its slave, *Leptothorax curvispinosus*. Ecology. 1957;**38**:449-456. DOI: 10.2307/1929889

[102] Passera L. Le cycle évolutif de la Fourmi *Plagiolepis pygmaea* Latr. (Hym. Form. Formicidae). Insectes Sociaux. 1963;**10**:59-70. DOI: 10.1007/BF02223522

[103] Passera L. Biologie de la reproduction chez *Plagiolepis pygmaea* Latreille et ses deux parasites sociaux *Plagiolepis grassei* Le Masne et Passera et *Plagiolepis xene* Stärcke (Hymenoptera, Formicidae). Annales des Sciences Naturelles – Zoologie. 1969;**11**:327-482

[104] Sanders C. The biology of carpenter ants in New Brunswick. Canadian Entomologist. 1964;**96**:894-909. DOI: 10.4039/Ent96894-6

[105] Sanders C. Seasonal and daily activity patterns of carpenter ants (*Camponotus spp.* ) in northwestern Ontario (Hymenoptera: Formicidae). Canadian Entomologist. 1972; **104**:1681-1687. DOI: 10.4039/Ent1041681-11

[106] Delage B. Recherches sur les fourmis moissonneuses du Bassin Aquitain: écologie et biologie. Bulletin Biologique de la France et de la Belgique. 1968;**102**:315-367

[107] Benois A. Évolution du couvain et cycle annuel de *Camponotus vagus* Scop (=*pubescens* Fabr.) (Hymenoptera, Formicidae) dans la région d'Antibes. Annales de Zoologie Ecologie Animale. 1972;**4**:325-351

[108] Dartigues D, Passera L. Polymorphisme larvaire et chronologie de l'apparition des castes femelles chez *Camponotus aethiops* Latreille (Hymenoptera, Formicidae). Bulletin de la Société Zoologique de France. 1979;**104**:197-207

[109] Cammaerts M-C. Étude démographique annuelle des sociétés de *Myrmica rubra* L. des environs de Bruxelles. Insectes Sociaux. 1977;**24**:147-161. DOI: 10.1007/BF02227168

[110] Ichinose K. Annual life cycle of *Paratrechina flavipes* (Hymenoptera, Formicidae) in the Tomakomai experiment Forest, southern Hokkaido. Kontyû. 1987;**55**:9-20

[111] Plateaux L. Dynamique des sociétés de la fourmi *Leptothorax nylanderi* (Förster). Biology and Ecology (Méditerranée). 1980;**7**:195-196

[112] Plateaux L. Comparaison des cycles saisonniers, des durées des sociétés et des productions des trois espèces de fourmis *Leptothorax* (*Myrafant* ) du groupe *nylanderi*. Actes des Colloques Insectes Sociaux. 1986;**3**:221-234

[113] Tohmé H, Tohmé G. La fondation de la société, le cycle biologique et ses variations chez *Messor syriacus* (Santschi) (Hym. Formicoïdea). Comptes Rendus de l'Académie des Sciences. Paris, Series D. 1980;**290**:1377-1379

[114] Tohmé G, Tohmé H. Le cycle biologique de *Messor incorruptus* (Ruzsky) (Hym. Formicoïdea). Corrélation entre l'espace disponible pour cette société de Fourmis et la productivité de celle-ci. Comptes Rendus de l'Académie des Sciences. Paris, Series III. 1982;**294**:115-118

[115] Cerdan P, Délye G. La fondation et les premières années du développement de la société de *Messor barbarus* (L.) (Hymenoptera, Formicidae). Comptes Rendus de l'Académie des Sciences. Paris. 1987;**305**:31-34

[116] Cerdan P, Délye G. L'hivernage et la ponte des femelles fondatrices de *Messor barbarus* (L.) (Hymenoptera Formicidae). Comptes Rendus de l'Académie des Sciences. Paris. 1990;**310**:231-236

[117] Schlick-Steiner BC, Steiner FM, Stauffer C, Buschinger A. Life history traits of a European *Messor harvester* ant. Insectes Sociaux. 2005;**52**:360-365. DOI: 10.1007/s00040-005-0819-8

[118] Meudec M. Cycle biologique de *Tapinoma erraticum* (Formicidae dolichoderinae). Annales de la Société Entomologique de France. 1973;**9**:381-389

[119] Mintzer A. Colony foundation and pleometrosis in *Camponotus* (Hymenoptera: Formicidae). Pan-Pacific Entomologist. 1979;**55**:81-89

[120] Brian M. The stable winter population structure in species of *Myrmica*. Journal of Animal Ecology. 1950;**19**:119-123. DOI: http://www.jstor.org/stable/1522

[121] Brian M. Ant culture for laboratory experiment. Entomologist's Monthly Magazine. 1951;**87**:134-136

[122] Brian M. Caste determination in a Myrmicine ant. Experientia. 1951;**7**:182-183. DOI: 10.1007/BF02148904

[123] Holmquist A. Notes on the life history and habits of the mound-building ant, *Formica ulkei* Emery. Ecology. 1928;**9**:70-87. DOI: 10.2307/1929544

[124] Holmquist A. Studies in arthropod hibernation. II. The hibernation of the ant, *Formica ulkei* Emery. Physiological Zoology. 1928;**1**:325-357. DOI: http://www.jstor.org/stable/30151051

[125] Scherba G. Reproduction, nest orientation and population structure of an aggregation of mound nests of *Formica ulkei* Emery ('Formicidae'). Insectes Sociaux. 1958;**5**:201-213. DOI: 10.1007/BF02224070

[126] MacKay E, MacKay W. Biology of the thatching ant *Formica haemorrhoidalis* Emery (Hymenoptera: Formicidae). Pan-Pacific Entomologist. 1984;**60**:79-87

[127] Kondoh M. Bioeconomic studies on the colony of an ant species, *Formica japonica* Motschulsky, 1: Nest structure and seasonal change of the colony members. Japanese Journal of Ecology. 1968;**18**:124-133. DOI: 10.18960/seitai.18.3_124

[128] Imamura S. Observations on the hibernation of a polydomous ant, *Formica* (*Formica*) *yessensis* Forel. Journal of the Faculty of Science, Hokkaido University. Series VI, Zoology. 1974;**19**:438-444. DOI: http://hdl.handle.net/2115/27569

[129] Gösswald K, Bier K. Untersuchungen zur Kastendetermination in der Gattung *Formica*. 3. Die Kastendetermination von Formica rufa rufo-pratensis minor Gossw. Insectes Sociaux. 1954;**1**:229-246. DOI: 10.1007/BF02222948

[130] Gösswald K, Bier K. Untersuchungen zur Kastendetermination in der Gattung *Formica*. 4. Physiologische Weisellosigkeit als Voraussetzung der Aufzucht von Geschlechtstieren im polygynen Volk. Insectes Sociaux. 1954;**1**:305-318. DOI: 10.1007/BF02329615

[131] Otto D. Die Roten Waldameisen. Die Neue Brehm-Bücherei. Vol. 293, Wittenberg, Lutherstadt: A. Ziemsen Verlag; 1962. 151 p

[132] Schmidt G. Steuerung der Kastenbildung und Geschlechtsregulation im Waldameisenstaat. In: Schmidt GH, editor. Sozialpolymorphismus bei Insekten. Stuttgart: Wiss. Verlag; 1974. pp. 404-512

[133] Cagniant H. Cycle biologique de la fourmi *Cataglyphis cursor* Fonscolombe, Hymenopteres, Formicidae. Vie et milieu. Série C, Biologie terrestre. 1977;**26**:277-281

[134] Cagniant H. La parthénogenèse thélytoque et arrhénotoque chez la fourmi *Cataglyphis cursor* Fonsc. (Hym., Form.). Cycle biologique en élevage des colonies avec reine et des colonies sans reine. Insectes Sociaux. 1979;**26**:51-60. DOI: 10.1007/BF02283912

[135] Cagniant H. La parthénogenèse thélytoque et arrhénotoque des ouvrières de la fourmi *Cataglyphis cursor* Fonscolombe. Etude en élevage de la productivité de sociétés avec reine et de sociétés sans reine. Insectes Sociaux. 1980;**27**:157-174

[136] Dlusskiy G, Soyunov O, Zabelin S. Murav'i Turkmenistana [Ants of Turkmenistan]. Ashkhabad, Ylym. 1989. 273 pp. (in Russian)

[137] Marikovskiy P. Murav'i Pustyn' Semirech'ya [Ants of the Deserts of Semirechie]. Alma-Ata: Nauka Kazakhskoy SSR; 1979. 263 pp. (in Russian)

[138] Kannowski P. Colony populations of two species of *Dolichoderus* (Hymenoptera: Formicidae). Annals of the Entomological Society of America. 1967;**60**:1246-1252. DOI: 10.1093/aesa/60.6.1246

[139] Torossian C. Recherches Sur la biologie et l'éthologie de *Dolichoderus quadripunctatus* (Hym. Form. Dolichoderidae). III. Étude expérimentale de l'action des principaux facteurs climatiques. Bulletin de la Société d' Histoire Naturelle de Toulouse. 1967;**103**:447-490

[140] Torossian C. Recherches sur la biologie et l'éthologie de *Dolichoderus quadripunctatus* (L.) Hym. Formicoidea Dolichoderidae. I. Étude des populations dans leur milieu naturel. Insectes Sociaux. 1967;**14**:105-121. DOI: 10.1007/BF02223262

[141] Cole Jr A. A brief account of aestivation and overwintering of the occident ant, *Pogonomyrmex occidentalis* Cresson, in Idaho. Canadian Entomologist. 1934;**66**:193-198. DOI: 10.4039/Ent66193-9

[142] Lavigne R. Bionomics and nest structure of Pogonomyrmex occidentalis (Hymenoptera: Formicidae). Annals of the Entomological Society of America. 1969;**62**:1166-1175. DOI: 10.1093/aesa/62.5.1166

[143] Hart L, Tschinkel W. A seasonal natural history of the ant, Odontomachus brunneus. Insectes Sociaux. 2012;**59**:45-54. DOI: 10.1007/s00040-011-0186-6

[144] Kipyatkov V, Lopatina E. Temperature and photoperiodic control of seasonal life cycles in ants (Hymenoptera, Formicidae). 1. Exogenous-heterodynamic species. Entomologicheskoe Obozrenie. 2003;**82**:801-819 (in Russian with English summary)

[145] Kipyatkov V. The role of endogenous rhythms in the regulation of annual cycles of development in ants (Hymenoptera, Formicidae). Entomologicheskoe Obozrenie. 1994;**73**:540-553 (in Russian with English summary, English translation in: Entomological Review. 1995;**74**:1-15)

[146] Lopatina E, Kipyatkov V. The influence of temperature on brood development in the incipient colonies of the ants *Camponotus herculeanus* (L.) and *Camponotus xerxes* Forel (Hymenoptera, Formicidae). In: Kipyatkov V, editor. Proceedings of the Colloquia on Social Insects. Vol. 2. Russian-speaking Section of the IUSSI, Socium: St. Petersburg; 1993. pp. 61-74

[147] Lopatina E, Kipyatkov V. The influence of daily thermoperiods on the duration of seasonal cycle of development in the ants *Myrmica rubra* L. and *M. ruginodis* Nyl. In: Kipyatkov V, editor. Proceedings of the Colloquia on Social Insects. Vol. 3-4. Russian-speaking Section of the IUSSI, Socium: St. Petersburg; 1997. pp. 207-216

[148] Lopatina E. Effect of daily thermoperiods on the duration of individual development in ants (Hymenoptera, Formicidae). Entomol. Obozr. 2003;**82**:537-547 (in Russian with English summary, English translation in: Entomological Review. 2004;**83**:1092-1101)

[149] Kipyatkov V, Lopatina E. Experimental study of seasonal cycle of rapid brood production in the ants *Myrmica rubra* L. and *M. ruginodis* Nyl. From two different latitudes. In: Kipyatkov V, editor. Proceedings of the Colloquia on Social Insects. Vol. 3-4. Russian-speaking Section of the IUSSI, Socium: St. Petersburg; 1997. pp. 195-206

[150] Kipyatkov V, Lopatina E. Seasonal cycle and winter diapause induction in ants of the genus *Myrmica* in the polar circle region. In: Kipyatkov V, editor. Proceedings of the Colloquia on Social Insects. Vol. 3-4. Russian-speaking Section of the IUSSI, Socium: St. Petersburg; 1997. pp. 277-286

[151] Kipyatkov V, Shenderova S. Endogenous rhythms in reproductive activity of red wood ant queens (*Formica rufa* group). Zoologicheskii Zhurnal. 1990;**69**:40-52 (in Russian with English summary)

[152] Bruniquel S. Observation Sur la biologie d' *Aphaenogaster subterranean* Latr. (Formicoidea, Myrmicidae). Etude experimentale du preferendum thermique. Bulletin de la Société d'histoire Naturelle de Toulouse. 1978;**114**:160-171

[153] Hölldobler B. Temperaturunabhängige rhythmische Erscheinungen bei Rossameisenkolonien (*Camponotus ligniperda* Latr. und *Camponotus herculeanus* L.) (Hym. Form.). Insectes Sociaux. 1961;**8**:13-22. DOI: 10.1007/BF02332768

[154] Soulié J. Action des facteurs du milieu sur le déclenchement et la rupture de l'état d'hibernation chez *Crematogaster scutellaris* Ol. (Hymenoptera: Formicoidea). Comptes Rendus des Seances de la Societe de Biologie. Paris. 1955;**149**:806-808

[155] Soulié J. Facteurs du milieu agissant sur l'activité des colonies de récolte chez la fourmi *Cremastogaster scutellaris* Ol (Hymenoptera, Formicoidea). Insectes Sociaux. 1955;**2**:173-177. DOI: 10.1007/BF02224102

[156] Soulié J. Le déclenchement et la rupture de l'état d'hibernation chez *Crematogaster scutellaris* Ol. (Hymenoptera: Formicoidea). Insectes Sociaux. 1956;**3**:431-438. DOI: 10.1007/BF02225763

[157] Soulié J. Quelques notes sur l'hibernation chez la fourmi *Crematogaster scutellaris* Ol. et chez une espèce voisine *Crematogaster auberti* Em. (Hymenoptera: Formicoidea). Insectes Sociaux. 1957;**4**:365-373. DOI: 10.1007/BF02224156

[158] Billen J. Stratification in the nest of the slave-making ant *Formica sanguinea* Latreille, 1798 (Hymenoptera, Formicidae). Annales de la Société royale zoologique de Belgique. 1984;**114**:215-225

[159] Dreyer W. The effect of hibernation and seasonal variation of temperature on the respiratory exchange of *Formica ulkei* Emery. Physiological Zoology. 1932;**5**:301-331. DOI: http://www.jstor.org/stable/30152792

[160] Brian M. Oviposition by workers of the ant *Myrmica*. Physiologia Comparata et Oecologia. 1953;**3**:25-36

[161] Brian M. Studies of caste differentiation in *Myrmica rubra* L. 5. Social conditions affecting early larval differentiation. Insectes Sociaux. 1962;**9**:296-310. DOI: 10.1007/BF02229629

[162] Brian M, Hibble J. Studies of caste differentiation in *Myrmica rubra* L. 7. Caste bias, queen age and influence. Insectes Sociaux. 1964;**11**:223-238. DOI: 10.1007/BF02222675

[163] Kipyatkov V. Studies of photoperiodic reaction in the ant *Myrmica rubra* L. (Hymenoptera, Formicidae). I. Main parameters of the reaction. Entomol. Obozr. 1974;**53**:535-545 (in Russian with English summary)

[164] Kipyatkov V. Ecology of photoperiodism in the ant *Myrmica rubra* L. (Hymenoptera, Formicidae). 1. Seasonal changes in the photoperiodic reaction. Entomol. Obozr. 1979; **58**:490-499 (in Russian with English summary, English translation in: Entomological Review. 1979;**58**:10-19)

[165] Kipyatkov V, Shenderova S. The effects of temperature and photoperiod on egg laying and productivity in red wood ant queens (Hymenoptera, Formicidae, *Formica rufa* group). Entomol. Obozr. 1991;**70**:36-46 (in Russian with English summary; English translation in: Entomological Review. 1991;**70**:13-24)

[166] Kipyatkov V. Studies of photoperiodic reaction in the ant *Myrmica rubra*. 4. Photoperiodic reactivation. Zoologicheskii Zhurnal. 1977;**56**:60-71 (in Russian with English summary)

[167] Hand L. The effect of photoperiod by the red ant, *Myrmica rubra*. Entomologia Experimentalis et Applicata. 1983;**34**:169-173. DOI: 10.1111/j.1570-7458.1983.tb03313.x

[168] Brian M. The importance of daylength and queens for larval care by young workers of the ant *Myrmica rubra* L. Physiological Entomology. 1986;**11**:239-249. DOI: 10.1111/j.1365-3032.1986.tb00411.x

[169] Kipyatkov V, Lopatina E. Temperature and photoperiodic control of Diapause induction in the ant *Lepisiota Semenovi* (Hymenoptera, Formicidae) from Turkmenistan. Journal of Evolutionary Biochemistry and Physiology. 2009;**45**:238-245. DOI: 10.1134/S0022093009020066

[170] Kipyatkov V, Lopatina E. Social regulation of larval diapause induction and termination by the worker ants of three species of the genus *Myrmica* Latreille (Hymenoptera, Formicidae). Entomol. Obozr. 1999;**78**:804-814 (in Russian with English summary; English translation in: Entomological Review. 2000;**79**:1138-1144)

[171] Kipyatkov V, Lopatina E. Vestigial photoperiodic response in subarctic *Myrmica* ants. Current Science. 2000;**79**:97-98

[172] Kipyatkov V. Studies of photoperiodic reaction in the ant *Myrmica rubra* L. (Hymenoptera, Formicidae). 5. Perception of photoperiodic information by an ant colony. Entomol. Obozr. 1976;**55**:777-789 (in Russian with English summary)

[173] Kipyatkov V. Myrmica prepares for winter. Science in the USSR. 1988;**1**:76-83

[174] Kipyatkov V, Lopatina E. Social regulation of larval development and diapause by the worker ants. In: Les Insectes Sociaux. In: 12th Congress of the International Union for the Study of Social Insects, Paris, Sorbonne, August 21-27, 1994. 1994. p. 317

[175] Tauber M, Tauber C, Masaki S. Seasonal Adaptations of Insects. New York, Oxford: Oxford University Press; 1986. XV+411 p

[176] Danks H. Insect Dormancy: An Ecological Perspective. Ottawa: Biological Survey of Canada (Terrestrial Arthropods); 1987. IX+439 p

[177] Danks H. Key themes in the study of seasonal adaptations in insects II. Life-cycle patterns. Applied Entomology and Zoology. 2006;**41**:1-13. DOI: 10.1303/aez.2006.1

[178] Elmes G. The social biology of *Myrmica* ants. Actes des Colloques Insectes Sociaux. 1991;**7**:17-34

[179] Elmes G, Thomas J. Die Biologie und Ökologie der Ameisen Gattung *Myrmica*. In: Geiger W, editor. Tagfalter und ihre Lebensraume: Arten. Gefahrdung, Schutz, Basle: Schweizerischer Bund für Naturschutz; 1987. pp. 404-409

Chapter 3

# Foraging and Predatory Activities of Ants

Ganesh Gathalkar and Avalokiteswar Sen

Additional information is available at the end of the chapter

http://dx.doi.org/10.5772/intechopen.78011

**Abstract**

Ants are a ubiquitous component of insect biodiversity and well known for its eusocial behavior. They are active foragers, scavengers, and predators that are prevalent in the vicinity of several plantations and crops. They (workers) prey on many insect species and feed on nectar exudates from plants as well as sticky secretions produced by Homopteran and Lepidopteran insects. As ferocious foragers with an aggressive attacking habit (e.g., *Oecophylla smaragdina*), they have often been used as biological control agents against various crop pests. However, some economically important insect species like the wild silkworm, *Antheraea mylitta*, are also affected by these foragers, namely, *O. smaragdina*, *Myrmicaria brunnea*, *Monomorium destructor*, *Monomorium minutum*, etc., which leads to the loss in crop outcome. In addition, some of them are known to destroy several plant species including domesticated fruit trees, particularly at the seedling stage. In this chapter, the foraging habit and the predation biology of these foragers are explored, in which the sequence of attack, their interactions, and invasion caused are discussed. It may also serve as a primary source of information on the foraging and its invasive impact, which may help to protect and/or take counteractive actions against the foragers which are harmful to commercial cultivations.

**Keywords:** aggressive predator, biological invasions, crop damage, foraging behavior, Tasar culture

## 1. Introduction

Ants (Hymenoptera: Formicidae) are eusocial cosmopolitan insects with about 13,262 species and 1941 subspecies, classified into 333 genera and 17 subfamilies [1]. They live in diverse habitats with diverse feeding habits and association with other species, in particular, plants and insects [2]. They form various colonies that consist of few to millions of individuals, living in small natural cavities to highly organized vast territories. Colonies comprise castes of sterile,

wingless females such as workers, as well as soldiers and other specialized groups [3, 4]. These ant colonies consist of fertile males, i.e., "drones," and one or few fertile females, i.e., "queens" [3], working together for the colony [5, 6]. Ants have colonized almost every landmass and may form about 15–25% of the terrestrial animal biomass [7]. Their social organization includes the ability to modify habitats and defend themselves. Ants form symbiotic associations with other organisms including other ant species, other insects, plants, etc. Their long coevolution with other species allowed them to enter into such mimetic, commensal, parasitic, or mutualistic relationships [2]. For example, in ant-fungus mutualism, both the species depend on each other for survival. The ant, *Allomerus decemarticulatus*, shows a three-way association with their host plant, *Hirtella physophora* (Chrysobalanaceae), and a sticky fungus helps to trap their insect prey [8]. They may, however, also be preyed upon by other animals as well, although their mimicry (myrmecomorphy), e.g., Batesian mimicry or Wasmannian mimicry (the mimic resembles its host to live within the same nest or structure) may reduce the risk of predation [9, 10]. In terms of their dietary requirements, most of the arboreal as well as some terrestrial taxa forage extensively on carbohydrate-rich plant secretions as well as insect exudates [2, 11]. Aphids and other hemipteran insects secrete a sweet liquid, i.e., honeydew, while feeding on plants. The sugars present in the honeydew are a high-energy food source [12]. Sometimes, the aphids secrete the honeydew for the ants so as to keep their predators away from them. Similarly, ants also tend to mealybugs to harvest their honeydew. Moreover, the myrmecophilous (ant-loving) larvae of the butterfly family Lycaenidae are driven by the ants. The larvae secrete the honeydew from their glands when the ants massage them, while some of them produce sounds and vibrations that are perceived by the ants [13].

As the ants are associated with another organism, they play a significant role in the insect ecosystem. During foraging, they feed on the plant cell sap and the honeydew produced by the other insects. However, they also feed on other insects to complete their food demand. As active foragers, they feed and affect several other commercially important insect species [2, 14–16], including the silkworm, *A. mylitta*, which affects the overall silk production [17, 18].

## 2. Biology and behavior of ants

### 2.1. Ants as biological control agent

As predators, ants are important in biological pest control efforts as their prey includes a range of insect species [15, 16]. Based on their foraging habits, the predatory ants can be classified as specialists or generalists [19]. Most of the species are scavengers where they prey on smaller organisms, as well as insect eggs. However, specialist ants do not seem to be significant in biological control measures, although some of them may have an impact on certain specific pests [20]. The generalist ant predators include those that are recognized as important in biological control [15, 16]. Most of the invasive ants are usually habitat generalists that allow them to invade and establish in undisturbed habitats [21]. Indigenous generalist predators have been controlling pests on crops since the dawn of agriculture, and the Chinese have used ant nests in citrus orchards to monitor the pest population [22]. It is now well documented that ants prey on eggs as well as larvae of numerous pest species in many different countries and habitats [20, 23]. The weaver ant, *O. smaragdina*, is a well-known predator which is used as a biological control agent against various agricultural pest species [20, 24, 25]. Similarly, the small red ant, *Formica rufa* (Linnaeus), is also known to kill many different defoliating pests in European forests [26]. Thus, the predacious generalist ants affect the behavior of prey directly and depress the size of potential pest populations [20, 27, 28]. Although, numerous insects possess generalized defense mechanisms, namely, flight, jumping, or dropping off the plant when vulnerable to attack by their enemies, but these may not be effective against ants that forage at different levels of the ecosystem [29]. The size and other physical attributes also aid in the mechanism of prey defense [27]. In addition, some ants are important in pollination, soil improvement, nutrient cycling, etc. [30]. In contrast, some feed on plants and may act as vectors of some plant diseases, while their attack may also be responsible for causing skin irritation in human beings, domestic animals, and other beneficial organisms [31, 32].

In the Tasar silkworm ecosystem, the worker ants of species such as *Oecophylla smaragdina* (Fabricius) (Hymenoptera: Formicidae), *Myrmicaria brunnea* (Saunders) (Hymenoptera: Formicidae), and some *Monomorium* species are frequent foragers which are harmful to the wild silkworm, *Antheraea mylitta*, resulting in losses in wild silk production [17, 18, 33]. The arboreal nature and highly aggressive predatory habit of these species of ants coupled with their extensive foraging on Tasar host plants (e.g., *Terminalia* sp.) often poses a severe risk in Tasar sericulture. Despite the knowledge of relative damage potential of predatory ants in the Tasar silkworm ecosystem, no systematic studies have been reported. Thus, to better understand the foraging and predatory behavior of these ant species on *A. mylitta*, a survey was undertaken in the Tasar rearing fields in Vidarbha, Maharashtra, India [18, 34, 35]. Furthermore, based on the symptoms of attack and predation on *A. mylitta*, the loss was also assessed.

### 2.2. Tasar silkworm (*Antheraea mylitta*)

The tropical silkworm, *Antheraea mylitta* (Drury) (Lepidoptera: Saturniidae), produces an excellent wild variety of silk, popularly known as "*Tasar* or *Kosa silk*," cultivated traditionally and commercially in India (see Gathalkar and Barsagade [18] for the lifecycle of *A. mylitta*) [36, 37]. The larvae of *A. mylitta* primarily feed on *Terminalia tomentosa* and *T. arjuna* besides several other secondary food plants [36, 37]. Tasar silkworm culture uplifts the socioeconomic status and provides a livelihood security to the stakeholders who are mainly the tribal folks [18, 31, 38]. The rearing of the Tasar silkworm is entirely wild, primarily in forests where it is exposed to various parasites and predators as well as to fungal, bacterial, and viral infections, thereby affecting the sericultural economics and the socioeconomic framework of tribal rearers/farmers [36, 39, 40].

### 2.3. Field conditions

The Tasar rearing sites of Bhandara (Lat. 21.059972, Long. 79.686987), Chandrapur (Lat. 20.399291, Long. 79.539701), and Gadchiroli (Lat. 20.508963, Long. 79.984988) and adjoining districts of Vidarbha in Maharashtra, India, were surveyed for studies on foraging and predatory behavior of ants in the Tasar ecosystem during the years 2014–2016. The climatic conditions of Tasar rearing zones were also recorded with the temperature ranging in between 35.5 ± 0.3 and 38.4 ± 0.2°C during the period of the first crop (June-August), 31.8 ± 0.2 and 33.4 ± 0.3°C in the second crop (August-November), and 17.4 ± 0.4 and 21.2 ± 0.3°C in third crop (November-February). The relative humidity was between 87.2 ± 0.2, 90.8 ± 0.6, and 77.2 ± 0.6% during the first to the third crops, whereas the average rainfall was about 362 ± 0.9, 196 ± 0.6, and 39 ± 0.5 mm during the first, second, and third crop, respectively.

### 2.4. Foraging turns to predation

The social Asian weaver ant, *Oecophylla smaragdina* (Fabricius), can be recognized by its nest building behavior. The workers are very active and fierce, and they are a serious predator of several insect species. They are a very common terrestrial as well as arboreal attacking forager, and, consequently, several studies have been conducted on the foraging behavior of various ant species, including *O. smaragdina* [30]. The highly aggressive feeding habit of *O. smaragdina* in the Tasar ecosystem is a challenge to the Tasar rearers where they attack the early larval instars of the Tasar silkworm, *Antheraea mylitta* [36, 37, 39], thereby affecting overall Tasar silk production. Similarly, the workers of *Myrmicaria brunnea* forages on the ground, leaves, and tree trunks [41, 42] including the Tasar host plants, namely., *T. tomentosa* and *T. arjuna* [18, 34, 43]. They are very aggressive predators and attack the larval stages of *A. mylitta* [34, 44]. While the ant species, namely, *Pheidologeton diversus* (Jerdon) and *Monomorium minutum* (Mayr), are documented as a predator of both Tasar and Muga silkworms [17, 18, 35], *Monomorium minimum* (Buckley) and *Pheidole* sp. attack the temperate Tasar silkworm, *Antheraea proylei* (Jolly) [17], and also *A. mylitta* [43]. Similarly, the ant *Tapinoma melanocephalum* (Fabricius) attacks the pupae and adults of the Muga silkworm [17, 33]. *Polyrhachis bicolor* (Smith) typically drags the spinning larvae in a group [45]. The ant species, namely, *Tetraponera rufonigra* (Jerdon), *Camponotus compressus* (Fabricius), and *Oecophylla smaragdina* (Fabricius), are very frequent foragers in the Tasar rearing fields [17, 18, 33, 34].

### 2.5. Predatory ants and their invasion in Tasar culture

There are numbers of colonies of predatory ants in the rearing fields of *A. mylitta* (D) in the forest zone of Bhandara, Gadchiroli, Chandrapur, and Gondia districts of Vidarbha, Maharashtra, India. The predatory attack by these predatory ants is very aggressive on the first to the third instar larvae of *A. mylitta* as well as during molting eventually leading to mortality. Their frequent bites on the larval integument and subsequent tearing with its sharp mandibles lead to death of the larvae [18]. Similarly, the small predatory ants (e.g., *Monomorium* sp.) also attack the pupa of *A. mylitta* [35]. However, the predation biology of these ants under field conditions is poorly known. Therefore, in the present chapter, the predation biology of these predatory and highly active foragers is discussed to unveil the risk of predation potential of these species besides the usual foraging habits of the ants.

#### 2.5.1. *Oecophylla smaragdina*

*Oecophylla smaragdina* is a very common forager attacking the early (first to third) larval instars of Tasar silkworm, and sometimes it attacks the fourth and fifth instar larvae as well, resulting in massive larval mortality [33, 36, 37]. The life cycle of *O. smaragdina* passes through egg, larval, pupal, and adult stages, and the nest exhibits division of labor with workers (reserve force, defenders, and nurses) and reproductive stages (male and female) [46]. The queen produces hundreds of eggs per day, and the worker population in the colony may total 500,000 offspring from a single queen [47]. The main criteria for separating castes are due to a reproductive capability which distinguishes the workers from the alates (or reproductive), and the males separate from gynes or females within the reproductive caste [48]. The worker ants are responsible for constructing their nest with the leaves of the host plant that is glued together by its larval silk. The workers are dimorphic, namely, major and minor forms, where the major workers are involved in the foraging and nest construction activity, and the minor workers remain in and around the nest, where they are involved in the maintenance of the colony and caring of the queen. In addition, the minor workers hold the larvae during weaving and nest building [49, 50]. *O. smaragdina* shows an extensive foraging for carbohydrate-rich plant secretions as well as insect exudates [2, 11]. Its bite on the human skin is painful due to the toxin sprayed on the wound from the tip of the gaster (e.g., *O. longinoda*) [49, 51]. Due to its far-reaching foraging habits and highly aggressive predatory behavior, *O. smaragdina* is being used as a biological control agent against major pests of economically important crops including many arthropods, acarid, isopod, myriapod, collembolan, termite, beetle, bark lice, and lepidopteran species and annelids like earthworms [20, 24, 52–54]. It can be used against the mango leafhoppers, thrips, fruit flies, tip borers, scale bugs, and mealy bug [55, 56].

##### 2.5.1.1. *Predatory behavior of O. smaragdina (worker)*

The sequence of attack: *O. smaragdina* (workers) follow the moving larvae and catch the larval appendages like hairs and setae with their sharp mandibles which leads to swelling, paralysis, and later the death of the larvae. Initially, the larva is captured by a single or few workers. Also, as a result of pricking of the integument and subsequent oozing of the hemolymph, the nearby ants are attracted. Often, they also carry the young larvae of *A. mylitta* to their nest (**Figure 1**).

##### 2.5.1.2. *Damage caused by O. smaragdina*

The attack of *O. smaragdina* is very aggressive; initially, one or very few predators attack the host larva followed by other ants in the vicinity. The ants tear the larva with their strong mandibles, which leads to oozing out of hemolymph and eventually causing larval mortality. The powerful mandibles of *O. smaragdina* are responsible for the painful bite besides irritation caused by the mandibular secretions [2]. The occurrence and subsequent invasion of *A. mylitta* by *O. smaragdina* also depend on abiotic factors like temperature, relative humidity, and rainfall. The attack of *O. smaragdina* on *A. mylitta* results in 4–5% loss in Tasar sericulture [18, 34].

**Figure 1.** Predation of *A. mylitta* by *O. smaragdina* showing (a) colony of *O. smaragdina* on Tasar host plant (*T. tomentosa*) and (b and c) predatory attack on early larval instars of *A. mylitta*.

**Figure 2.** The predatory attack of ant *M. brunnea* (a) on a small insect, (b) on the larva of *A. mylitta* on the tree trunk of host plant, and (c) on the larva of *A. mylitta* on the Tasar ground (field) (source: Gathalkar and Barsagade [43]).

### 2.5.2. *Myrmicaria brunnea*

*Myrmicaria brunnea* of subfamily Myrmicinae has a distinctively curved abdomen and two spines on the metathorax. Workers are chestnut brown with shining mandibles. The genus *Myrmicaria* is predominantly a honeydew feeder and scavenger, which builds underground nests. Some species of *Myrmicaria* are highly predatory, foraging in groups and moving in a sinuous path with widely opened antennae [57]. It is a dominant predator of many insect species, including the larvae of *A. mylitta* besides earthworms.

The workers of *M. brunnea* were found to forage on the Tasar host plants, *T. tomentosa* and *T. arjuna*. It builds ground nests under the Tasar host plant, and it shows terrestrial as well as arboreal scrounging propensity. They suck the cell sap from the leaves (ventral side) of Tasar host plant, and during sap sucking, they also attack small insect species (**Figure 2(a)**) as well as larvae of Tasar silkworm (**Figure 2(b)** and **(c)**). Initially, the Tasar larvae are captured by a few workers and subsequently pricked, thereby attracting other workers nearby the site of attack.

The workers are highly aggressive, cut their prey into small pieces, and later on transport them to their ground nest, and sometimes the whole prey is also transported to the nest (**Figure 2(c)**). Sometimes, these predatory ants carry their prey to their ground nest either after cutting into small pieces or the whole prey including the fourth/fifth instar larvae of *A. mylitta*.

Being aggressive, the predation activity of *M. brunnea* and weaver ant, *O. smaragdina*, shows remarkable similarities in Tasar rearing also [34, 58–60]. The ant, *M. brunnea* (Saunders), has a geniculate type of the antenna which is characteristic of aculeate Hymenoptera [55, 56, 61]. A ball-like scape at the base region present in the ants, *Diacamma* sp. and *Camponotus japonicus* Mayr [55, 61], is also observed in *M. brunnea*. The pedicel in *M. brunnea* is long and broad with an imbricate surface and covered with patches of sensilla, similar to *C. japonicus, C. sericeus* [56, 61], and *C. compressus* [33].

The mouthparts of the ant species are well developed and adapted for grasping and feeding on the host species. The mandibles in *M. brunnea* are potent tools for prey catching, fighting, digging, wood-scraping, grooming, brood care, and trophallaxis [2, 62]. The abundance of *M. brunnea* in Tasar rearing fields is a serious issue, which affects the total Tasar silk production [18]. Predation by *M. brunnea* was also recorded on Muga silkworm, *A. assamensis* [17].

#### *2.5.2.1. Feeding behavior (Myrmicaria brunnea)*

The attack and feeding pattern of this ant are very aggressive. Initially, one or very few ants attack the larva of *A. mylitta*, and, subsequently, other members of the colony join the group for feeding (**Figure 2**) [43]. As feeding progresses, the ants tear the host larva with their robust and sharp mandibles due to which hemolymph oozes followed by the complete destruction of the prey (Supp. Info. video clip 1 (can be viewed at https://youtu.be/q8WfVBLLlvA). The ant, *M. brunnea*, usually attacks the early instars of Tasar silkworm; we also observed them to attack the fourth and fifth instar larvae (**Figure 2(c)**). During feeding, the larvae of *A. mylitta* often fall to the ground which are then attacked by these ants. They may consume the whole prey at the site, or they drag their prey to their ground nest (**Figure 2(c)**) (Supp. Info. video clip 2 (can be viewed at: https://youtu.be/JsbbiWeZOw0)). During the predatory attack, the Tasar host larvae try to escape, but the intensity of injuries and constant biting by the ants make the larva defenseless, resulting in complete larval invasion and eventual death.

#### *2.5.2.2. Damage (crop loss)*

The mean percent of larval mortality of *A. mylitta* due to the attack by *M. brunnea* (workers) was calculated, and the year-wise mortality was about 3–5% of total crop damage (**Figure 3**) [18, 43].

### *2.5.3. Monomorium sp.*

The myrmicine genus, *Monomorium*, includes the small-sized ants, reddish-brown in color, and belongs to the family Formicidae. There are about 358 species in which the genus *Monomorium* includes 27 subspecies [63]. They represent one of the most influential groups of ants due to its abundant diversity and intra-morphological and biological variability [64]. Of these, *Monomorium pharaonis* (Linnaeus), *Monomorium destructor* (Jerdon), and *Monomorium floricola* (Jerdon) are well-known household pests [65]. As a predator of various pest species, they are often used in pest management programs. The predatory habit of ants has a major influence in many habitats [66, 67]. Thus, some ants are biologically essential

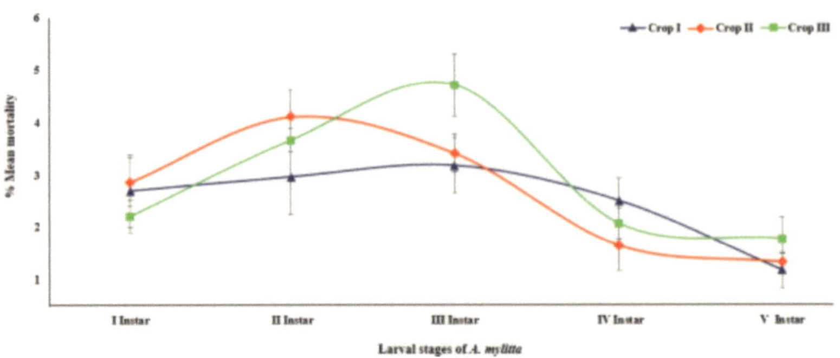


**Figure 3.** Percent mortality of *A. mylitta* (larva) by *M. brunnea* (source: Gathalkar 2014 [43]).

**Figure 4.** Predation of *Antheraea mylitta* showing (a) larval attack by *Monomorium minimum* (transporting the first instar larva), (b) *M. destructor* on the cocoon (pupa), and (c) damaged pupa of *A. mylitta* (source: Gathalkar and Barsagade [35]).

for the pollination, predation, scavenging, soil improvement, nutrient cycling, as well as plant dispersal [30, 41, 68]. However, in the Tasar ecosystem, the workers of *Monomorium* species including *M. destructor* and *M. minimum* attack the early larval instars (first to third) of *A. mylitta* (**Figure 4(a)**) besides entering into the cocoon by making holes and feeding on the pupa (**Figure 4(b)** and **(c)**). They attack silkworms during resting and molting on trees, while the pupae, adult, and eggs are primarily affected at grainage.

The ants around households feed on any food available [69]. *Monomorium destructor* is a small ant, which exhibits polymorphism and varies in size from 1.8 to 3.5 mm [70]. These are common household pests, and the foragers are slow in finding food compared with other tramp ants [71]. They are a minor component of the ant fauna with *M. floricola* (Jerdon), *O. smaragdina*, *Crematogaster* sp., and *Paratrechina longicornis* (Latreille) being the most common ants [23]. *Monomorium destructor* forms large polygyne colonies [69], where they form their nest predominantly in trees in hollow twigs and branches and the soil in tropical regions as well [69]. Different foraging patterns employed by the different ant species [72] are in a proportion of foragers whose feeding on liquid food demonstrates high trophallaxis rates [73]. The foraging workers of *Monomorium* sp. are passive movers unlike the erratic foragers from the *Tapinorna* or *Paratrechina* genera [74]. Similarly, *Pheidole* sp. is the major predators of *Alabama argillacea* eggs [75].

In urban populations, ants also cause frequent problems where they destroy the esthetic and other products of human consumption [2, 71]. Occasionally, they also act as vectors of various plant diseases. The attack of some ant species is quite painful to domestic animals as well as human beings [31, 32]. However, these ant species can also be used as an ecological indicator, to assess the ecological status regarding species diversity and the impact of invasive species [76].

#### *2.5.3.1. Behavioral studies*

Feeding habits and prey distraction (field invasion): the ants *Monomorium minimum* and *M. destructor* have their terrestrial nests on the Tasar host plants, including *Terminalia tomentosa* and *T. arjuna*, and can be recognized by their conspicuous trail [35]. While foraging, the worker ants attack several larvae of *A. mylitta* as well as pupae, thereby affecting a broad range of host stages (**Figure 4**). Their attack on late instar disturbs the entire spinning process as well as larval development. Due to feeding on the larvae as well as pores made on the cocoon shell, the quality as well as the overall production of raw silk is affected. Some of the ants also carry their prey to their colony. Despite their small size, they are capable of attacking and preying upon much larger larvae of *A. mylitta* (**Figure 4(a)**) (Sup. Info. 3: https://youtu.be/jSycX5tAuMg). During predation, the first instar larva of *A. mylitta* tries to escape many times, but the mandibular grips of *Monomorium* make Tasar larvae attempt to escape futileness [35]. Also, a single ant can also drag the whole first instar larva of the silkworm. Sometimes, they also feed on the late instar larva of *A. mylitta*, which may either be previously damaged by another predator, dead or diseased. Quite often, the damage is severe, and care should be taken during rearing of Tasar silkworm.

#### *2.5.3.2. Damage by Monomorium*

The destruction of larvae of the Tasar silkworm by ant predators is severe, and the damage caused to the cocoons due to the pores results in broken silk threads rendering in a loss to the sericulture industry. It also causes a drop in the silkworm population in subsequent generations. The crop-wise mortality is estimated to be between 2 and 4% [18].

### 2.6. Role of sensory organs in the foraging habits of ants

The antenna of *O. smaragdina* consists of scape, pedicel, and flagellomeres in all castes, with 10 flagellomeres observed in males and 11 in females (workers and queen) [77]. Various types of antennal sensilla have previously been reported in the ants, *Lasius fuliginosus* (Latreille) [78] and *Diacamma* sp. [79, 80]. In *O. smaragdina* (worker), the scape is covered with polygonal cuticular plates (which form the cuticular micro-sculpturing) along with sensilla trichoidea (ST-I and ST-II). In addition, there are three types of sensilla basiconica (SB-I, SB-II, and SB-III) (**Figure 5**). Moreover, STC and ST are present densely on the flagellar segments, while the last two flagellar segments reveal the presence of SB and sensilla ampullacea (SA). The sensilla coeloconica (SC) and SA are intense on the middle surface of the terminal flagellar segments (**Figure 5(k)**). Thus, the presence of these types of sensilla in *O. smaragdina* is similar to sensilla reported in other Hymenopteran species [78–82].

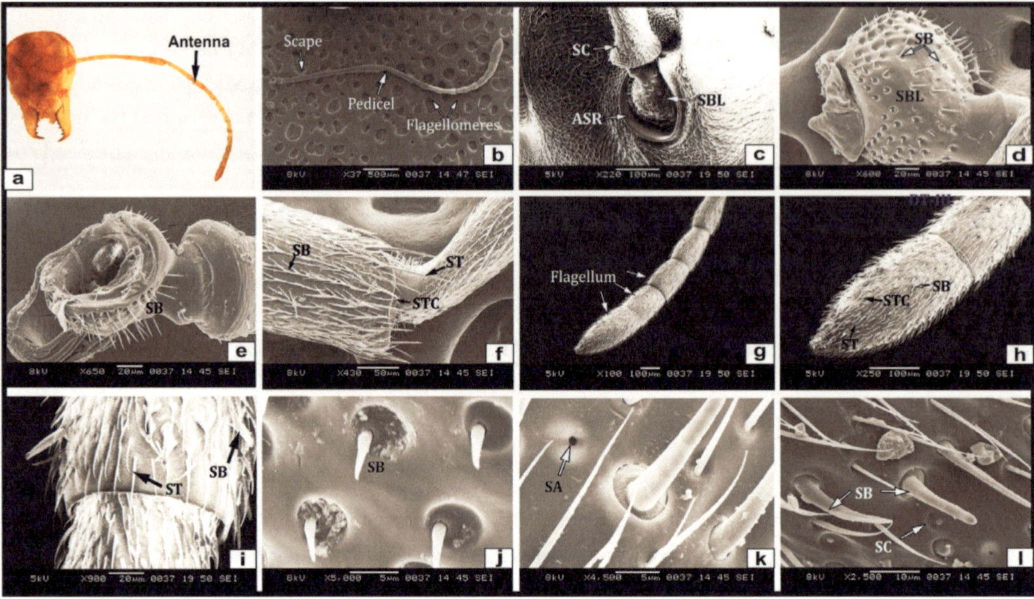


**Figure 5.** Structure of the antenna of *O. smaragdina* (worker) showing (a) light microscopic view of the head and antenna; SEM micrographs showing (b) the antenna of worker, (c) basal region of antenna, (d) magnified view of scape ball, and (e) scape galina; and magnified view of scape showing shaft base with sensilla, (f) sensilla present on pedicel, (g) funiculus, (h) detailed view funiculus with sensilla, (i) intersegment of the antenna and sensilla, (j) sensilla basiconica, (k) sensilla ampullacea along with basiconica, and (l) basiconica and coeloconica sensilla. *Abbreviations*: SC, scape; SBL, scape ball; ASR, antennal sclerotic ring; ST, sensilla trichoidea; STC, sensilla trichoidea curvata; SB, sensilla basiconica; SC, sensilla coeloconica; and SA, sensilla ampullacea (source: Gathalkar 2014 [43]).

In most of the ant species, the mouthparts are adapted for grasping and feeding on the prey [83, 84]. Paul et al. [85] reported that gustatory sensilla are situated on the lower pair of jaws in the ant. The mandibles in *O. smaragdina* and *M. brunnea* are potent tools for prey catching, fighting, digging, seed crushing, wood-scraping, grooming, brood care, and trophallaxis [2, 86]. There are two types of sensilla trichoidea (ST-I, ST-II) and STC present on the labrum. The ST is on the dorsal surface (DT-I, DT-II, and DT-III in the figure) and on the ventral surface into VT-I and VT-II types (**Figure 6**). The sensilla DT-I is present in the marginal area of the dorsal region of mandibles. The morphology of sensilla in males is similar to that of female except for difference in size (**Figure 6**). On the dorsal side of the mandibles, trichoid sensilla are densely distributed, whereas, on the ventral side, sensilla basiconica predominates. SB is also found in worker mandibles. The labium shows the presence of sensilla ST-I, ST-II, and STC (**Figure 6(m)** and **(n)**). The maxilla is endowed with sensilla trichoidea (ST-I and ST-II) and STC, while the inner surface of maxilla is filled with sensillary fold along with the ST (**Figure 6(o)** and **(p)**). The trichoid sensilla and small peg-like sensilla basiconica on the dorsal and ventral surface of mandibles in dragonfly were reported as mechanoreceptors and chemoreceptors, respectively [87, 88]. Similar sensilla trichoidea and sensilla basiconica observed on the mandible of *C. compressus* [89] and also observed in *O. smaragdina* might be performing a similar function as mechano- and chemoreceptors.

http://dx.doi.org/10.5772/intechopen.78011

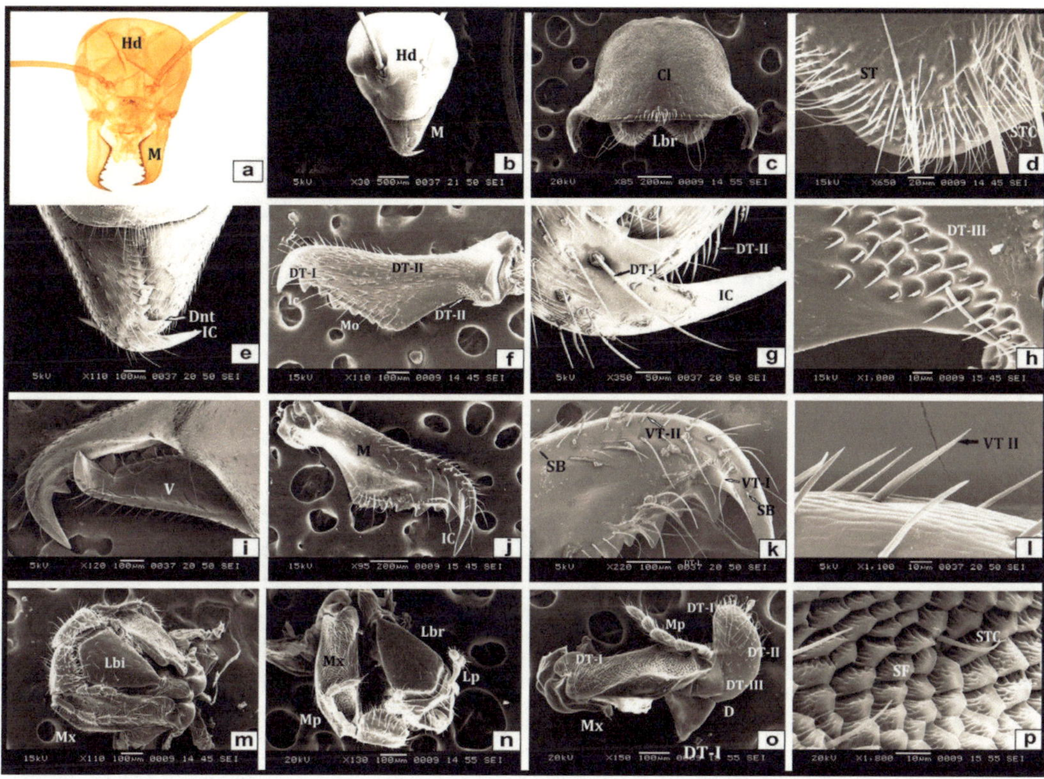


**Figure 6.** Mouthparts of *Oecophylla smaragdina* (worker) showing (a) light microscopic view of the head with the mandible, SEM view of (b) the head with mouthparts, (c) closed view of the labrum with clypeus, (d) detailed view of the labrum with sensilla, (e) mandible view in detail, (f) the mandible in dorsal view, (g) the mandible with the teeth and sensilla, (h) sensilla trichoidea (DT), (i) closed position of mandibles showing both the dorsal and ventral sides of mandible, (j) the mandible in ventral view, (k) the mandible with the teeth and sensilla; (l) sensilla trichoidea (VT), (m) complex structure of the labium and maxilla; (n) the labium with maxilla and palpi; (o) maxilla with palpi and sensilla, and (p) detailed inner maxillary folds. *Abbreviations*: H, head; M, mandible; cl, clypeus; Lbr, labrum; IC, incisor; MO, molar; D, dorsal side; DT, dorsal trichoid sensilla, Dnt, dentations; Lbi, labium; mx, maxilla; Mp, maxillary palp; SF, sensory folds; V, ventral side; VT, ventral trichoid sensilla (source: Gathalkar 2014 [43]).

Sensilla on the maxillary and labial palpi were characteristically different in their morphology. Sensilla with a bifid curved porous tip suggest a chemosensory function [77]. The present work, therefore, confirms the presence of various types of sensilla on mandibles in worker caste of the ant which play a crucial role in the predatory and feeding behavior of *O. smaragdina*.

The geniculate antenna of *M. brunnea* is elbow shaped, consisting of a scape, pedicel, and five flagellomeres (**Figure 7(a)**) [90]. The scape is covered with polygonal cuticular plates with three types of sensilla basiconica (SB-I, SB-II, and SB-III) (**Figure 7(b)**). The entire surface of the elongated shaft of the scape is also covered with the polygonal cuticular plates as well as sensilla trichoidea ST-I and ST-II. Trichoid sensilla are present throughout the surface of the pedicel in worker ants (**Figure 7(b)**). The flagellum (**Figure 7(c)** and **(d)**) is covered with sensilla trichoidea curvata (STC) and sensilla trichoidea (ST) and two types of sensilla basiconica. The SC is concentrated on terminal flagellar segments at middorsal position.

Scanning electron micrographs reveal the diversity and density within each of the four basic types of antennal sensilla of *M. brunnea*, namely, the SB, ST, STC, and SC. Similar sensilla were reported on *C. compressus* [91] and other Hymenoptera [78, 79, 81, 82, 92]. Sensilla trichoidea located on the antennae of *M. brunnea* at the pedicel region have also been reported in other species [79, 82]. The SB on the antennae exhibits a similar morphological structure to previously studied ant species and may function as contact gustatory sensilla [80, 82, 93]. The antennal sensilla basiconica (SB) of fire ants, *Solenopsis invicta*, is also known to function as a contact chemoreceptor [94, 95]. Nakanishi et al. [82] categorized two types of trichoid sensilla along with the sensilla trichoidea curvata in *C. japonicus* which does not always respond to

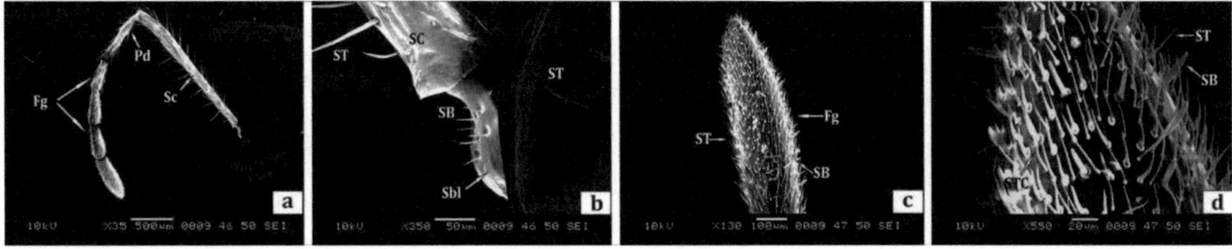


**Figure 7.** Scanning electron microscopic structure of the antenna of *M. brunnea* showing (a) close view of the antenna with sensilla, (b) antennal socket with ball and scape with sensilla, (c) flagellum, and (d) sensilla present on the tip of the flagellum. *Abbreviations*: Sc, scape; Sbl, scape ball; Fg, flagellum; Pd, pedicel; ST, sensilla trichoidea; STC, sensilla trichoidea curvata; SB, sensilla basiconica (source: Gathalkar and Barsagade [90]).

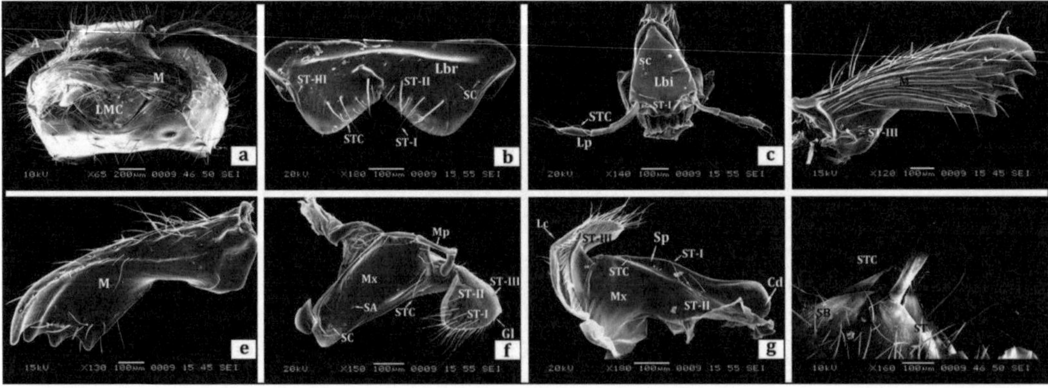


**Figure 8.** SEM structure of the mouthparts of *M. brunnea*, showing (a) front view of the mouth with arrangements of mouthparts, (b) labrum, (c) labium, (d) dorsal view of the mandible, (e) ventral view of the mandible, (f) dorsal view of the maxilla, (g) ventral view of the mandible, and (h) sting apparatus. *Abbreviations*: A, antenna; Lbr, labrum; Lbi, labium; M, mandible; Mx, maxilla; LMC, labio-maxillary complex; Mp, maxillary palp; Lp, labial palp; cd, cardo; Sp, stipes; Lc, lacinia; Gl, galea; ST, sensilla trichoidea; STC, sensilla trichoidea curvata; SB, sensilla basiconica; SA, sensilla ampullacea; and SC, sensilla coeloconica (source: Gathalkar and Barsagade [90]).

http://dx.doi.org/10.5772/intechopen.78011

stimulation by alarm pheromones [92, 96]. Thus, these may have a similar function in *M. brunnea* also. The STC in *M. brunnea* resembles those in other ant species [82, 97], which may perform as contact chemosensilla [82, 98].

In *M. brunnea* [90], ultrastructural studies reveal the presence of three types of sensilla, namely, ST, STC, and SC, with three distinct types of trichoid sensilla, namely, ST-I, ST-II, and ST-III (**Figure 8(a–h)**). Additionally, on the labial palp, ST and STC are observed (**Figure 8(c)**). On the mandibles, three types of ST, SB, and SC are observed. The sensilla ST-I is present on the marginal area of the dorsal region of mandibles, while SC is observed on the upper peripheral region (**Figure 8(d)** and **(e)**). In several Myrmicinae, moderately stipulated sting apparatus, which may be spatula shaped as observed in *M. opaciventris*, are well described [99, 100].

During predation, these ants deposit venom into the prey's cuticle by wagging the bent gaster [57].

The furcula, a wishbone-shaped sclerite whose ventral arms are flexible, is attached to the base of the sting, causing the aculeus to pitch, roll, and yaw in probing for a sting site [101].

## 3. Conclusion

The foraging behavior of various ant species may be harmful or beneficial depending on the host species. In Tasar sericulture, we find ants like *O. smaragdina* and *M. brunnea* which are highly aggressive predators, as well as *Monomorium* sp. With an understanding of the population dynamics of these species, preventive measures can be adopted to prevent losses.

It also helps to develop future pest control strategies to minimize the loss of commercially important crops. The approaches necessary to bring down the losses in Tasar rearing sites due to these predatory ants need to be reevaluated, and in this regard, the possibility of using semichemicals offers a suitable alternative.

## Acknowledgements

We gratefully acknowledge the suggestions made by Prof. D. D. Barsagade, RTM Nagpur University, Nagpur, which has improved the manuscript. We also thank the Central Silk Board, Government of India, for allowing us to access the Tasar rearing fields in Vidarbha and the ICAR-Indian Agricultural Research Institute (IARI), Pusa, New Delhi, for the identification of the insect specimens.

## Conflict of interest

We do not have any conflict of interest.

## Author details

Ganesh Gathalkar* and Avalokiteswar Sen

*Address all correspondence to: g.gathalkar@ncl.res.in

Division of Organic Chemistry, Laboratory of Entomology, CSIR-National Chemical Laboratory, Pune, Maharashtra, India

## References

[1] Bolton B. An Online Catalog of the Ants of the World [Internet]. 2017. http://www.antcat.org [Accessed: November 10, 2017]

[2] Hölldobler B, Wilson EO. The Ants. Cambridge, MA: Belknap Press of Harvard University; 1990. p. 732

[3] Singh R. Elements of Entomology. Meerut, UP, India: Rastogi Publications; 2006. p. 284

[4] Fisher BL, Bolton B. Ants of Africa and Madagascar: A Guide to the Genera. Oakland, California, USA: University of California Press; 2016. p. 24. ISBN 9780520290891

[5] Oster GF, Wilson EO. Caste and Ecology in the Social Insects. Princeton: Princeton University Press; 1978

[6] Flannery T. A Natural History of the Planet. New York, USA: Grove/Atlantic, Inc; 2011. p. 79

[7] Schultz TR. In search of ant ancestors. Proceedings of the National Academy of Sciences. 2000;**97**(26):14028-14029. DOI: 10.1073/pnas.011513798

[8] Dejean A, Solano PJ, Ayroles J, Corbara B, Orivel J. Arboreal ants build traps to capture prey. Nature. 2005;**434**(7036):973. DOI: 10.1038/434973a

[9] Reiskind J. Ant-mimicry in Panamanian clubionid and salticid spiders (Araneae: Clubionidae, Salticidae). Biotropica. 1977;**9**(1):1-8. DOI: 10.2307/2387854

[10] Cushing PE. Myrmecomorphy and myrmecophily in spiders: A review. The Florida Entomologist. 1997;**80**(2):165-193. DOI: 10.2307/3495552

[11] Davidson DW. The role of resource imbalances in the evolutionary ecology of tropical arboreal ants. Biological Journal of the Linnean Society. 1997;**61**:153-181

[12] Styrsky JD, Eubanks MD. Ecological consequences of interactions between ants and honeydew-producing insects. Proceedings of the Biological Sciences. 2007;**274**(1607):151-164. DOI: 10.1098/rspb.2006.3701

[13] DeVries PJ. Singing caterpillars, ants and symbiosis. Scientific American. 1992;**267**(4):76-82. DOI: 10.1038/scientificamerican1092-76

[14] Gosswald K. Die Waldameise. Band 2. Die Waldemeise im Okosystem Wald, ihr Nutzen und ihre Hege. Wiesbaden: Aula-Verlag; 1990. 510p

[15] Petal J. The role of ants in ecosystems. In: Brian MV, editor. Production Ecology of Ants and Termites. IBP 13:293-325. Cambridge: Cambridge University Press; 1978

[16] Risch SJ, Carroll CR. Effect of a keystone predaceous ant, *Solenopsis geminata*, on arthropods in a tropical agroecosystem. Ecology. 1982;**63**:1979-1983

[17] Negi BK, Siddiqui AA, Sengupta AK. Insect pests of Muga silkworms and their management. Indian Silk. 1993;**32**:37-38

[18] Gathalkar GB, Barsagade. Parasites-predators: Their occurrence and invasive impact on the tropical Tasar silkworm *Antheraea mylitta* (Drury) in the zone of central India. Current Science. 2016;**111**(10):1643-1657. DOI: 10.18520/cs/v111/i10/1649-1657

[19] Wilson EO. Some ecological characteristics of ants in New Guinea rain forests. Ecology. 1959;**40**:437-447

[20] Way MJ, Khoo KC. Colony dispersion and nesting habits of the ants, *Dolichoderus thoracicus* and *Oecophylla smaragdina* (Hymenoptera: Formicidae), in relation to their success as biological control agents on cocoa. Bulletin of Entomological Research. 1991; **81**:341-350

[21] Passera L. Characteristics of tramp species. In: Williams D, editor. Exotic Ants. Boulder: Westview Press; 1994. pp. 22-43

[22] Symondson WOC, Sunderland KD, Greenstone MH. Can generalist predators be effective biocontrol agents? Annual Review of Entomology. 2002;**47**:561-594

[23] Way MJ, Cammell ME, Bolton B, Kanagaratnam P. Ants (Hymenoptera: Formicidae) as egg predators of coconut pests, especially in relation to biological control of the coconut caterpillar, *Opisina arenosella* Walker (Lepidoptera: Xyloryctidae), in Sri Lanka. Bulletin of Entomological Research. 1989;**79**:219-233

[24] Paulson GS, Akre RD. Evaluating the effectiveness of ants as biological control agents of pear psylla (Homoptera, Psyllidae). Journal of Economic Entomology. 1992;**85**:70-73

[25] Way MJ, Javier G, Heong KL. The role of ants, especially the fire ant, *Solenopsis geminata* (Hymenoptera: Formicidae), in the biological control of tropical upland rice pests. Bulletin of Entomological Research. 2002;**92**:431-437

[26] Pascovici VD. Especes du groupe *Formica rufa* L. de Roumanie et leur utilisation dans la lutte contre les ravageurs forestieres. Proceedings of the Meeting of the Working Groups on *Formica rufa* and Vertebrate Predators of Insects. Bulletin SROP—International Organization for Biological Control of Noxious Animals and Plants, West Palaearctic Regional Section, 1979; 2: 111-34. Varenna, Italy: West Palearctic Regional Section of the 10BC

[27] Way MJ, Khoo KC. Role of ants in pest management. Annual Review of Entomology. 1992;**37**:479-503

[28] Rico-Gray V, Oliveira PS. The Ecology and Evolution of Ant-Plant Interactions. Chicago: University of Chicago Press; 2007

[29] Heads PA, Lawton JH. Bracken, ants and extrafloral nectaries. III. How insect herbivores avoid ant predation. Ecological Entomology. 1985;**10**:29-42

[30] Gotwald WH. The beneficial economic role of ants. In: Vinson SB, editor. Economic Impact and Control of Social Insects. New York: Praeger; 1986. pp. 290-313

[31] Vinson SB, editor. Economic Impact and Control of Social Insects. New York: Praeger Publishers; 1986. 421p

[32] Goddard J, de Shazo R. Fire Ant Attacks on Humans and Animals. In: Encyclopedia of Pest Management. Boca Raton, Florida, USA: CRC Press; 2004. DOI: 10.1081/E-EPM-120024662

[33] Singh KC. Controlling the insect enemies of oak tasar silkworms. Indian Silk. 1991; **30**(7):19-23

[34] Gathalkar GB, Barsagade DD. Predation biology of weaver ant, *Oecophylla smaragdina* (Hymenoptera: Formicidae) in the field of tasar sericulture. Journal of Entomology and Zoology Studies. 2016;**4**(2):07-10

[35] Gathalkar GB. Barsagade DD. Predatory potential of two species of *Monomorium* on the developing stages of silkworm *Antheraea mylitta* (Drury) (Lepidoptera: Saturniidae). Entomon. 2017;**42**(3):227-234

[36] Jolly MS, Chaturvedi SM, Prasad SA. Survey of Tasar crops in India. Indian Journal of Sericulture. 1968;**1**:50-58

[37] Jolly MS, Sen SK, Sonwalkar TN, Prasad GK. Non-mulberry silks. Food and Agriculture Organization of United Nations, Service Bulletin. 1979;**29**:1-178

[38] Jolly MS. Package of practices for tropical tasar culture, Ranchi. Bombay: Central Tasar Research Station, (Central Silk Board); 1976. p. 32

[39] Singh RN, Thangavelu K. Parasites and Predators of Tasar silkworm, *Antheraea mylitta* and their control. Indian Silk. 1991;**29**(12):33-36

[40] Barsagade DD. Tropical Tasar Sericulture. In: Omkar, editor. Industrial Entomology. Singapore: Springer; 2017. DOI: 10.1007/978-981-10-3304-9_10

[41] Lach L, Parr CL, Abbott KL, editors. Ant Ecology. 1st ed. Oxford: Oxford University Press; 2010. 402p

[42] General DM, Alpert GD. A synoptic Review of the ant genera (Hymenoptera: Formicidae) of the Philippines. ZooKeys. 2012;**200**:1-111

[43] Gathalkar GB. Studies on the pests and predators of tropical Tasar silkworm *Antheraea mylitta* eco-race Bhandara (Lepidoptera: Saturniidae) [thesis]. Nagpur: RTM Nagpur University; 2014

[44] Barsagade DD, Gathalkar GB. *Myrmicaria brunnea* (Saunders) (Hymenoptera: Formicidae): A new predator in tasar sericulture. Proceedings of the 10th ANeT International Conference. Department of Zoology and Environmental Management, University of Kelaniya, Sri Lanka (Abstract); 2015. p. 53. URI: http://repository.kln.ac.lk/handle/123456789/10162

[45] Bidyapati L, Noamani MKR, Das PK. Pest complex of oak tasar. Indian Silk. 1994;**33**(3):44-48

[46] Hingston RWC. The habits of *Oecophylla smaragdina*. Proceeding of Entomological Society of London. 1927;**2**:90-94

[47] Hölldobler B, Wilson EO. Weaver ants—Social establishment and maintenance of territory. Science. 1977;**195**:900-902

[48] Wilson EO. The soldier of the ant, *Camponotus* (*Colobopsis*) *fraxinicola*, as a trophic caste. Psyche. 1974;**81**:182-188

[49] Weber NA. The functional significance of dimorphism in the African ant, *Oecophylla*. Ecology. 1949;**30**:397-400

[50] Ledoux A. Recherche sur la biologie de la fourmi fileuse (*Oecophylla longinoda*). Annales des Sciences Naturelles. 1950;**12**:313-461

[51] Vanderplank FL. The bionomics and ecology of the red tree ant *Oecophylla* sp., and its relationship to the coconut bug *Pseudotheraptus wayi* Brown (Coreidae). Journal of Animal Ecology. 1960;**29**:15-33

[52] Peng RK, Christian K, Gibb K. The effect of colony isolation of the predacious ant, *Oecophylla smaragdina* (F.) (Hymenoptera: Formicidae), on protection of cashew plantations from insect pests. International Journal of Pest Management. 1999;**45**:189-194

[53] Van Mele P, Cuc NTT. Evolution and status of *Oecophylla smaragdina* as a pest control agent in citrus in the Mekong Delta, Vietnam. International Journal of Pest Management. 2000;**46**(4):295-301

[54] Cerda X, Dejean A. Predation by ants on arthropods and other animals. In: Polidori C, editor. Predation in the Hymenoptera: An Evolutionary Perspective. Kerala: TransWorld Research Network; 2011. pp. 39-78

[55] Peng RK, Christian K. The weaver ant, *Oecophylla smaragdina* (Hymenoptera: Formicidae), an effective biological control agent of the red-banded thrips, *Selenothrips rubrocinctus* (Thysanoptera: Thripidae) in mango crops in the Northern Territory of Australia. International Journal of Pest Management. 2004;**50**:107-114

[56] Cole AC, Jones JW. A study of the weaver ant, *Oecophylla smaragdina* (Fab.). American Midland Naturalist. 1948;**39**:641-651

[57] Kenne M, Schatz B, Durand JL, Dejean A. Hunting strategy of a generalist ant species proposed as a biological control agent against termites. Entomologia Experimentalis et Applicata. 2000;**94**:31-40

[58] Peng R, Christian K, Reilly D. The effect of weaver ants *Oecophylla smaragdina* on the shoot borer *Hypsipyla robusta* on African mahoganies in Australia. Agricultural and Forest Entomology. 2011;**13**:165-171. DOI: 10.1111/j.1461-9563.2010.00514.x

[59] Offenberg J, Havanon S, Aksornkoae S, Macintosh DJ, Nielsen MG. Observations on the ecology of weaver ants *Oecophylla smaragdina* (Fabricius) in a Thai mangrove ecosystem and their effect on herbivory of *Rhizophora mucronata* Lam. Biotropica. 2004;**36**:344-351

[60] Tsuji K, Hasyim A, Harlion, Nakamura K. Asian weaver ants, *Oecophylla smaragdina,* and their repelling of pollinators. Ecological Research. 2004;**19**:669-673

[61] Peng RK, Christian K. The control efficacy of the weaver ant, *Oecophylla smaragdina* (Hymenoptera: Formicidae), on the mango leafhopper, *Idioscopus nitidulus* (Hemiptera: Cicadellidea) in mango orchards in the Northern Territory. International Journal of Pest Management. 2005;**51**:297-304

[62] Grogenberg W, Hölldobler B, Alpert GD. Jaws that snap: The mandible mechanism of the Mystrium. Journal of Insect Physiology. 1998;**44**:241-253

[63] Bolton B. An online catalog of the ants of the world [Internet]. 2017. http://www.antcat.org/catalog/429718 [Accessed: November 10, 2017]

[64] Aslam M, Shaheen FA, Ayyaz A. Management of *Callosobruchus chinensis* Linnaeus in stored chickpea through interspecific and intraspecific predation by ants. World Journal of Agricultural Sciences. 2006;**2**(1):85-89

[65] Williams DF. Control of the introduced pest Solenopsis invicta in the United States. In: Williams DF, editor. Exotic ants. Boulder, Colorado: Westview; 1994. pp. 282-291

[66] Wilson EO. The Insect Societies. Cambridge, MA: Belknap; 1971. 548p

[67] Carroll CR, Janzen DH. Ecology of foraging by ants. Annual Review of Ecological Systems. 1973;**4**(23):1-57

[68] Folgarait PJ. Ant biodiversity and its relationship to ecosystem functioning: A review. Biodiversity and Conservation. 1998;**7**:1221-1244

[69] Smith MR. Household-infesting ants of the eastern United States: Their recognition, biology, and economic importance. U.S. Department of Agriculture Technical Bulletin. 1965;**1326**:105

[70] Harris R, Abbott K, Barton K, Berry J, Don W, Gunawardana D, Lester P, Rees J, Stanley M, Sutherland A, Toft R. Invasive ant pest risk assessment project for Biosecurity New Zealand. Series of unpublished Landcare Research contract reports to Biosecurity, New Zealand 2005; BAH/35/2004-1

[71] Lee CY. Tropical household ants: Pest status, species diversity, foraging behaviour, and baiting studies. In: Jones, SC, Zhai, J, Robinson WH, editors. Proceedings of the 4th International Conference on Urban Pests; Virginia, Pocahontas Press; 2002. pp. 3-18

[72] Ayre GL. Problems in using the Lincoln Index for estimating the size of ant colonies (Hymenoptera: Formicidae). Journal of New York Entomological Society. 1962;**60**:159-167

[73] Stradling DJ. Food and feeding habits of ants. In: Brian MV, editor. Production Ecology of Ants and Termites. Cambridge: Cambridge University Press; 1978. pp. 81-106

[74] Edwards JP. The biology, importance and control of pharaoh's ant, *Monomorium pharaonis* (L.). In: Vinson SB, editor. Economic Impact and Control of Social Insects. New York: Praeger Publishers; 1986. pp. 257-271

[75] Gravena S, Pazetto JA. Predation and parasitism of cotton leafworm eggs, *Alabama argillacea* (Lep.: Noctuidae). Entomophaga. 1987;**32**:241-248

[76] Bharti H, Bharti M, Pfeiffer M. Ants as bioindicators of ecosystem health in Shivalik Mountains of Himalayas: Assessment of species diversity and invasive species. Asian Myrmecology. 2016;**8**:1-15

[77] Babu MJ, Ankolekar SM, Rajashekhar KP. Castes of the weaver ant *Oecophylla smaragdina* (Fabricius) differ in the organization of sensilla on their antennae and mouthparts. Current Science. 2011;**101**(6):1-10

[78] Dumpert K. Alarm stoffrezeptoren auf der Antenne von *Lasius fuliginosus* (Hymenoptera: Formicidae). *Zeitschrift für Vergleichende Physiologie*. 1972;**76**:403-425

[79] Okada Y, Miura T, Tsuji K. Morphological differences between sexes in the ponerine ant, *Diacamma* sp. (Formicidae: Ponerinae). Sociobiology. 2006;**48**:527-541

[80] Ozaki M, Wada-Katsumata A, Fujikawa K, Iwahasi M, Yokahari F, Satoji Y, Nishimurat YY. Ant nestmate and non nestmate discrimination by a chemosensory sensillum. Science. 2005;**309**:311-314

[81] Esslen J, Kaissling KE. Zahl-und Verteilung antennaler Sensillen bei der Honigbiene (*Apis mellifera* L.). Zoomorphology. 1976;**83**:227-251

[82] Nakanishi A, Nishino H, Watanabe H, Yokohari F, Nishikawa M. Sex-specific antennal sensory system in the ant *Camponotus japonicus*, structure and distribution of sensilla on the flagellum. Cell and Tissue Research. 2009;**338**:79-97

[83] Snodgrass RE. Principles of Insect Morphology. New York: Mc Graw Hill; 1935

[84] Chapman RF. The Insect Structure and Function. 4th ed. Cambridge: Cambridge University Press; 1998

[85] Paul JP, Flavio R, Hölldobler B. How do ants stick out their tongues? Journal of Morphology and Embryology. 2002;**254**:39-52

[86] Paul J. Mandible movements in ants. Comparative Biochemistry and Physiology. 2001; **13**(1):7-20

[87] Zacharuk RY. Ultrastructure and function of insect Chemosensilla. Annual Review of Entomology. 1980;**25**:27-47

[88] Wazalwar SV, Tembhare DB. Mouthparts sensilla in dragon fly, *Brachjythemes contaminata* (Fabricius) (Anosoptera: Libelllidae). Odonatologia. 1999;**28**(3):257-271

[89] Barsagade DD, Tembhare DB, Kadu SG. SEM structure of mandibular sensilla in the carpenter ant *Camponotus compressus* (Fabricius) (Formicidae: Hymenoptera). Halters. 2010;**2**(1):53-57

[90] Gathalkar GB, Barsagade DD. Cephalic microstructure and its role in predation biology of *Myrmicaria brunnea* on *Antheraea mylitta*. Journal of Applied Biology and Biotechnology. 2018;**6**(1):1-6. DOI: 10.7324JABB.2017.60101

[91] Barsagade DD, Tembhare DD, Kadu SG. Microscopic structure of antennal sensilla in the carpenter ant *Camponotus compressus* (Fabricius) (Formicidae: Hymenoptera). Asian Myrmecology. 2013;**5**:113-120

[92] Zacharuk RY. Antennae and sensilla. In: Kerkut GA, Gilbert LI, editors. Comprehensive Insect Physiology, Biochemistry and Pharmacology. Vol. 6. Nervous System. Oxford, UK: Sensory. Pergamon Press; 1985. pp. 1-69

[93] Mysore K, Shyamala BV, Rodrigues V. Morphological and developmental analysis of peripheral antennal chemosensory sensilla and central olfactory glomeruli in worker caste of *Camponotus compressus* (Fabricius, 1787). Arthropod Structure & Development. 2010;**39**:310-321

[94] Renthal RD. Sensory reception in fire ants, Imported Fire Ant Research and Management Project, Final report, Texas; 2003. pp. 1-3

[95] Renthal R, Velasqueza D, Olmosa D, Hamptona J, Wergin WP. Structure and distribution of antennal sensilla of the red imported fire ant. Micron. 2003;**34**:405-413

[96] Nakanishi A, Nishino H, Watanabe H, Yokohari F, Nishikawa M. Sex-specific antennal sensory system in the ant *Camponotus japonicus,* glomerular organizations of antennal lobes. Journal of Comparative Neurology. 2010;**518**:2186-2201

[97] Hashimoto Y. Unique features of sensilla on the antennae of Formicidae (Hymenoptera). Applied Entomology and Zoology. 1990;**25**:491-501

[98] Altner H, Prillinger L. Ultrastructure of invertebrate chemo-, thermo- and hygroreceptors and its functional significance. In: Bourne GH, Danielli JF, editors. International Review of Cytology. Academic Press; 1980. pp. 69-139

[99] Kugler C. Evolution of the sting apparatus in the myrmicine ants. Evolution. 1979;**33**: 117-130

[100] Kaib M, Dittebrand H. The poison gland of the ant *Myrmicaria eumenoides* and its role in recruitment communication. Chemoecology. 1990;**1**:3-11

[101] Kugler C. A comparative study of the myrmicine sting apparatus (Hymenoptera: Formicidae). Studia Entomologica. 1978;**20**:413-548

Chapter 4

# Leaf-Cutter Ants and Microbial Control

Raphael Vacchi Travaglini,
Alexsandro Santana Vieira, André Arnosti,
Roberto da Silva Camargo,
Luis Eduardo Pontes Stefanelli, Luiz Carlos Forti and
Maria Izabel Camargo-Mathias

Additional information is available at the end of the chapter

http://dx.doi.org/10.5772/intechopen.75134

**Abstract**

The attini tribe comprises fungusgrowing ants, such as the basal *Apterostigma* and other more specialized genera, including the higher attine and the ones that cut the fresh plant tissue (*Atta* and *Acromyrmex*), maintaining an obligatory mutualistic relation with the fungus *Leucoagaricus gongylophorus*, which serves as a food source for the ants. Leaf-cutter ants are considered agriculture pests and populate the soil, a rich environment, especially due to the presence of several microorganisms. Some of these microorganisms are natural enemies that may cause epizootics (quickly spreading opportunistic diseases). Such defence strategies include polyethism, that is, division of labor among the individuals. The older ants take on the responsibility of foraging, as their integument is harder and heavily sclerotized, serving as a protective barrier against pathogens (including bacteria and antagonistic fungi). The younger ants, whose metapleural glands synthetize important secretions to eliminate and control microorganisms that could attack the colony fungus garden and the immature (larvae and pupae), remain inside the colony cultivating symbiont fungi. The sum of the survival strategies of ants in general, including social immunity and nest-cleaning behavior, represents a barrier for the application of biological control programs, mainly microbial ones.

**Keywords:** epizootics, entomopathogenic, fungi

## 1. Origin

Leaf-cutter ants, classified into more than 250 species within 17 genera, are found exclusively in the American continent [1, 2]. The basal genera *Cyphomyrmex*, *Mycetophylax*, *Mycocepurus*,

*Myrmicocrypta, Apterostigma, Mycetosoritis* and *Mycetarotes* use dead vegetable matter and insect feces and carcasses as substrate for the fungi. The genus *Pseudoatta* comprises parasite species and do not have a worker caste. The genera *Trachymyrmex, Sericomyrmex, Attaichnus, Kalathomyrmex, Paramycetophylax, Acromyrmex* (quenquens) and *Atta* (sauvas) belong to the superior attini group.

Based on the nidification habits of these species, the lowland arid environments with open vegetation in South America are suggested to be the centers of diversification [3, 4]. However, some studies have reported that this diversification occurred in the wet environment of the Amazon Basin [5, 6]. Recent molecular studies on attini ants point to dry environments as a decisive factor in the symbiont fungus domestication [7].

## 2. Mutualistic interactions

Attini ants have maintained a mutualistic relationship with basidiomycete fungi, in an obligatory association for around 50 million years [8]. The interaction with the fungus is not restricted to feeding, once symbiotic associations between larvae and fungi in basal attini have been reported in the literature [9]. Based on morphometric analyses, *Sericomyrmex* and *Trachymyrmex* together form a distinct group from the other genera. The transition from the ancestral agricultural system toward the derived leaf-cutting habit is followed by remarkable changes in nest size and architecture, colony size and worker size and polymorphism. Considering *Sericomyrmex* and *Trachymyrmex* as possessing transitional habits, differently from species that cultivate fungus by using mostly non-plant items (insect feces and carcasses) as well as from typical leaf-cutters *Atta* and *Acromyrmex*. Studies detect correlations of nest traits with worker number and size of the fungus garden in the less conspicuous attini [10].

## 3. Agronomic importance

The genera *Atta* and *Acromyrmex* play an important role in protecting the soil structure by recycling organic matter; however, these ants have been reported to damage the aerial parts of cropped plants in a short period of time [11]. Several strategies have been developed to control these insects [12]. In general, to be efficient against leaf-cutter ants, toxic baits must contain active ingredients with retarded action [13]. In addition, the insecticide must present specific characteristics, such as being lethal in low concentrations, not killing the ants immediately when applied in high concentrations, being absorbed through ingestion and having slow action, being easily distributed throughout the colony, being highly degradable to avoid environmental impact and not having a repellent or deterrent action on the ants [14]. Thus, the search for new alternatives to control leaf-cutter ants is ongoing, and the use of plant extracts and microbial control has been proven to be promising strategies.

## 4. Biology and behavior

The colonies of *Atta* ants are founded by the queens, that visit the colony fungus gardens before the nuptial flight and place a small piece of fungal mycelium (Leucocoprinus) in their infrabuccal cavities [15]. After this, the queen burrows into the ground [16], lays the eggs and takes care of the brood until the first workers eclose, take over the fungus culture activities and help in the construction and expansion of the chambers, more specifically the garbage and symbiont ones [11].

Ants are holometabolous insects, that is, their life cycle includes three stages: larva, pupa and adult. They are organized into castes, which perform different functions, including the maintenance of the colony and brood care and feeding. The males and queens are part of the reproductive castes, ensuring the survival of the species.

Leaf-cutting ants are prevalent herbivores and dominant invertebrates in tropical forests: the volume of the soil occupied by a single 6-year-old nest of *Atta sexdens* weighs approximately 40,000 kg, and a colony with such dimensions can collect more than 5892 kg of leaves [17]. Therefore, *Atta* ants are considered important ecosystem engineers once they modify the soil composition and create regions with high concentrations of organic matter, causing the development of more demanding vegetal species in terms of nutrition and altering the flora composition [18]. These effects may remain up to 15 years after the death of the colony of nest abandonment [19].

The colonies present a broad behavioral repertoire, varying according to the morphology (polymorphism) and age (polyethism) of the workers [20–22]. The foraging behavior consists of cutting and transporting the vegetal substrate to the interior of the colony, where it is processed and incorporated in the symbiont fungus garden [5, 17]. It is the most intense and energy-demanding activity performed in the colony, requiring approximately 90% of the workers [23].

## 5. Microbial control

### 5.1. Entomopathogenic fungi

Entomopathogenic fungi (EF) have been proven a promising strategy to control leaf-cutter ants through microorganisms. These fungi present two phases in their biological cycle: anamorphic (vegetative) and reproductive, when they produce conidia, and, according to Sung et al. [24], they belong to the phylum Ascomicota and class Hypocreales (**Table 1**).

Considered one of the most efficient pathogens in microbial control, entomopathogenic fungi need a host to spread in the environment, and the literature has reported that they attach to the cuticle of pest insects through physical and chemical processes, producing chitinolytic enzymes and developing a penetration clamp from the apical region of the hyphae [25, 26]. When the host insect is debilitated (low immunity level), some opportunistic fungi can act as saprophytes, accelerating the insect's death process [27]. Most fungi penetrate the host through the integument [28, 29]. According to Quiroz [30], who

| | Family | |
|---|---|---|
| | **Clavicipitaceae** | **Cordycipitaceae** |
| Telemorphs | *Hypocrella, Metacordyceps* | Cordyceps s.str., |
| | *Regiocrella, Torrubiella* | *Torrubiella* |
| Anamorphs | *Aschersoni, Metarhizium,* | *Beauveria, Isaria,* |
| | | *Lecanicillium* |

**Table 1.** Some fungi.

studied *A. mexicana* ants, entomopathogenic fungi are one of the most important mortality agents for the queens. The author identified the following species: *Aspergillus parasiticus, Paecilomyces farinosus, Beauveria bassiana* and *Metarhizium anisopliae*. In Brazil, most studies have used *B. bassiana* and *M. anisopliae*. Alves and Sosa-Gomez [31] reported these fungi virulence in worker and reproductive castes of *Atta sexdens rubropilosa*. In forests of *Eucalyptus grandis*, promising results were obtained using *B. bassiana*, in baits to control *Acromyrmex* spp [32].

Studies by Kermarrec et al. [33], which lasted approximately 10 years, used several entomopathogenic fungi, applied on *Acromyrmex octospinosus*, popularly known as 'quenquem', and they showed that this insect is able to identify and isolate the pathogen, leading the authors to the conclusion that the social surveillance/immunity and altruism of the workers would be essential for microbial control. Fungi as *B. bassiana, M. anisopliae* and *Trichoderma viridae* were proven inefficient when used in the control of *Acromyrmex heyeri* ants [34].

Machado et al. [35] reported that ants of the genus *Acromyrmex* would present systems, probably olfactive, with the function of recognizing entomopathogenic fungi. Such functions would be the result of selective pressure that occurred during the evolution of this social insect, which nests on pathogen-abundant soils. This would be accompanied by behaviors related to defense strategies aimed at reducing the dissemination and the effects of entomopathogenic fungi on the colony [36].

In this sense, it is known that the ant antennae "analyze" different types of material, allowing the recognition of pathogenic elements that could harm the colony [37]. Kermarrec et al. [33] observed that the sensitivity of the antennae can be demonstrated by the fact that attractive baits containing entomopathogenic fungi and spores are not cut but placed away from the colony area by *Ac. octospinosus*. Studies using electroantennography demonstrated that the olfactive function of *Ac. octospinosus* would be dependent on neuroreceptors located in the antennae [36]. Morphological analyses using scanning electron microscopy showed differentiation in these sensorystructures in *Atta robusta* [38]. In a recent study, whose results serve as a model for ants in general, Slone et al. [39] analyzed *Harpegnathos saltator* and identified several subfamilies of olfactory receptors, demonstrating the complexity and sophistication of this sensory organ.

## 6. Other entomopathogenic microorganisms

In addition to entomopathogenic fungi, other organisms have been tested to control pest insects. Entomopathogenic nematodes (*Steinernema* and *Heterorhabditis*) were proven inefficient against urban ants [40] or caused only partial mortality to a more susceptible leaf-cutter species. These nematodes associate with entomopathogenic bacteria (*Xenorhabdus* and *Photorhabdus*) found in their digestive tract. Isolated and multiplied in aqueous media, *Acromyrmex subterraneus* were reported susceptible to the entomopathogenic bacterium *Photorhabdus temperate* K122, highly virulent when inoculated in worker abdomens [41]. Fungicide metabolites produced in the cultivation media (supernatant) have been used to control phytopathogenic fungi in field applications [42]; therefore, leaf-cutter ant control can be performed by targeting the symbiont fungus. The use of endophytic bacteria to overcome the morphological, mechanical and biochemical defenses of leaf-cutter ants aiming to contaminate the symbiont fungus [43], as well as the use of endophytic fungi [44], has been regarded as promising research lines; however, the assumptions are still hypothetical.

It is important to emphasize that the associations of the bacteria *Pseudonocardia, Streptomyces* and *Burkholderia* with the integument of leaf-cutter ants would represent a barrier for the microbial control success, once these bacteria secrete compounds to defend the host, occupying the niche where the pathogenic agent would settle [45, 46].

## 7. Use of synthetic chemicals

Once toxic substances are capable of overcoming insect immunity barriers (individually and collectively) [47, 48], the development research on strategies to impair the colony organization through immunosuppression, weakening the humoral system of the ants, is fundamental. Such strategies often include the use of chemical substances [49]. In this sense, subdoses of insecticides would aid the biological control; however, this would lead us to the use of granular baits containing a low concentration of the active ingredient. Nevertheless, the bait itself is highly specific to the target insect and, consequently, environmentally friendly, in contrast with other insecticides available in the market, and the demand for efficient and sustainable products is on the increase, considering the stricter requirements of regulatory agencies [50].

## 8. Capability to neutralize pathogens through gland secretions

Leaf-cutter microbial control has been subject to the same criticism directed toward plat extracts, that is, considering that the ants evolve in an environment where toxic plants and pathogenic microorganisms abound, probably the concentration is not the only issue to be taken into consideration for the success of these control methods [51–53]. The virulence

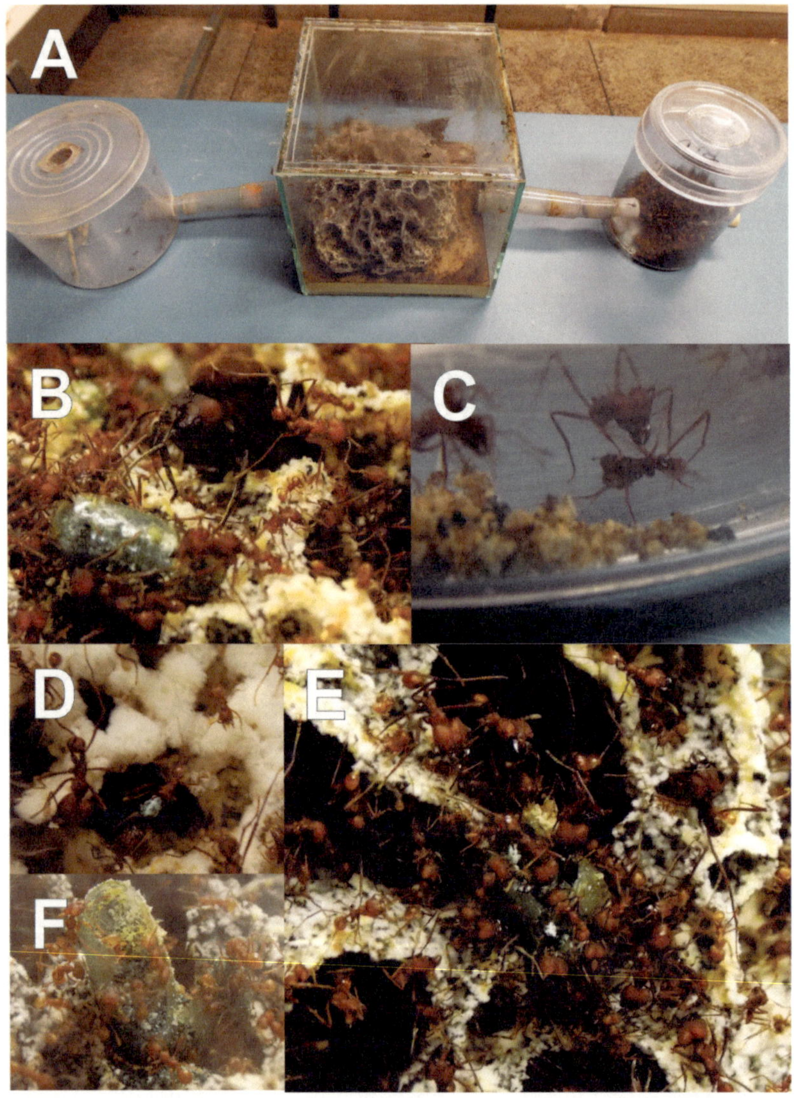


**Figure 1.** Previews with parted gelatin capsules containing *Trichoderma asperellus:* (A) colony kept in laboratory, (B) queen next to the capsule containing *T. asperellus,* (C) worker disposing the antagonist fungus in the waste chamber, (D) workers manipulating the fungus *T. asperellus,* (E and F) workers incorporating the capsule and content. (Source: Author).

of entomopathogenic fungi has been intensely investigated, especially the length of time needed to cause the pest population mortality; however, further investigation is needed on the forms of application and compatibility with adjuvants. Moreover, microbial control faces several natural barriers, including the capability of the ants to inhibit the germination of the conidia through secretions produced by their salivary glands (4-methyl 3-heptanone) and through their fecal fluid (chitinolytic enzymes) as well [54]. According to Pagnocca [55], in order to prevent the contamination of the symbiont fungus garden, the ants lick the foraged substrate surface, keeping microorganisms in the infrabuccal cavity to dispose them in the waste chamber. Thus, the infrabuccal cavity would function as a filter, preventing the entrance of solid particles into the worker body and in the nest as well [56, 57].

The metapleural glands represent another defense mechanism, secreting several substances, such as phenylacetic acid; 3-hydroxydecanoic acid; indoleacetic acid; and skatole [58], which produce acid secretion that inhibits the germination of some entomopathogenic fungi, protecting the cuticle against microbial infections [57]. Fernandez-Marin et al. [59] identified a higher level of cleanliness by the gland secretion in workers inoculated with *Paecilomyces lilacinus* conidia. This chemical strategy developed by the ants to protect the fungi present in the environment is maximized by the grooming habit, either individual or collective, definitely removing the pathogenic microorganism from the host's body, preventing adhesion and, consequently, the infection onset [60].

However, some authors reckon that the pathogen is transmitted to part of the population during grooming, possibly contaminating larvae (immature) and ultimately the queen [61, 62] (**Figure 1**).

## 9. Perspectives

Pest control must be thought of from a preventive point of view and, in the case of leaf-cutter ants, in the initial phase of colony establishment. The scientific scenario today allows a more thorough and accurate research [34].

- The use of advanced analysis tools, such as scanning, light and confocal microscopes to investigate what happens in the moment when the insect is infected by the fungus [63].
- Cutting-edge technology software to monitor the agronomic environment and the emergence of entomological radars [64] can be incorporated as control strategies, indicating the best moment for the technique application (nuptial flight or "revoada").
- Drones have been used in some sampling techniques for leaf-cutter ants [65].
- The use of microbial control agents in different formulations (conidia, blastospores or microscleroids) synergically applied along with adjuvants is a tendency in microbial control [66].

## 10. Final consideration

Overall, microbial control has been proven efficient to control some pest insects (*Cosmopolite sordidus, Mahanarva fimbriolata, Diaphorina citri,* among others), except the ones presenting social behavior [67]. Leaf-cutter ants are endowed with an accurate system to identify and fight microorganisms (antennae) and secret several antimicrobial compounds through different glands located throughout their bodies [68, 69]. In addition, the grooming behavior constantly performed by the workers makes microbial control almost impossible with the use of the currently available tools.

Various isolates of *Escovopsis sp.* were analyzed in order to understand whether the antagonism of these fungi involves natural products, as suggested by the genome filled with genes encoding mycotoxins and fungal cell wall degrading-enzymes [71], the greater knowledge of pathogens such as *Escovopsis sp.* and *Trichoderma sp.* can be a way to found potential control agent for the control of leafcutting ants [72], reported in experiments, including those of systematic fungal analyzes of the symbiotic fungal garden where antagonistic fungi manifest aggressive development in the presence of stresses in the colonies, even those maintained for a long period in the laboratory [73]. In the case of the antagonistic fungi is the possibility of a joint action between microorganisms for control. However, although there is potential in the laboratory, it will not necessarily be promising in the field, due to the necessity of successive applications and the longer time in relation to the chemical, which is already long (30 days), to obtain colony death [70].

## Author details

Raphael Vacchi Travaglini[1]*, Alexsandro Santana Vieira[2], André Arnosti[2], Roberto da Silva Camargo[1], Luis Eduardo Pontes Stefanelli[1], Luiz Carlos Forti[1] and Maria Izabel Camargo-Mathias[2]

*Address all correspondence to: raphaelvacchitravaglini@gmail.com

1 Department of Plant Protection, School of Agriculture, São Paulo State University-UNESP, Botucatu, SP, Brazil

2 Department of Cellular and Molecular Biology, Institute of Biosciences, São Paulo State University-UNESP, Rio Claro, SP, Brazil

## References

[1] Brandão CRF, Mayhé-Nunes AJ, Sanhudo CED. Taxonomia e filogenia das formigas-cortadeiras. In: Della-Lucia TMC, editor. Formigas cortadeiras: Da bioecologia ao manejo. 1st ed. Viçosa: UFV; 2011. pp. 28-48

[2] Sosa-Calvo J, Schultz T, Brandão CRF, Klingenberg C, Feitosa RM, Rabeling C, Bacci MJ, Lopes CT, Vasconcelos HL. Cyatta abscondita: Taxonomy, evolution, and natural history of a new fungus-farming ant genus from Brazil. PLoS One. 2013;**8**:e80498

[3] Fowler HG. Distribution patterns of Paraguayan leaf-cutting (Atta and Acromyrmex) (Formicidae: Attini). Studies on Neotropical Fauna and Environment. 1983;**18**:121-138

[4] Kusnezov N. Zoogeografia de las hormigas en Sudamerica. Acta Zoologica Lilloana (Argentina). 1963;**19**:25-186

[5] Weber NA. Gardening Ants, the Attines. Philadelphia: American Philosophical Society; 1972

[6] Mayhé-Nunes AJ, Jaffé K. On the biogeography of Attini (Hymenoptera: Formicidae). Ecotropicos. 1998;**11**(1):45-54

[7] Branstetter M, Jesovnik A, Sosa-Calvo J, Lloyd MW, Faircloth BG, Brady SG, Schultz TR. Dry habitats were crucibles of domestication in the evolution of agriculture. Proceedings Royal Society B. 2017;**284**:20170095. DOI: 10.1098/rspb.2017.0095

[8] Schultz TR, Brady SG. Major evolutionary transitions in ant agriculture. Proceedings of the National Academy of Sciences of the USA. 2008;**105**:5435-5440

[9] Ortiz G, Mathias MIC, Bueno OC. First evidence of an intimate symbiotic association between fungi and larvae in basal attine ants. Micron. 2012;**43**:263-268

[10] Leal IR, Silva PSD, Oliveira PS. Natural history and ecological correlates of fungus-growing ants (Formicidae: Attini) in the Neotropical Cerrado savanna. Annals of the Entomological Society of America. 2011;**104**:901-908

[11] Della-Lucia TMC. Formigas cortadeiras: da bioecologia ao manejo. 1st ed. Viçosa: UFV; 2011

[12] Boaretto MAC, Forti LC. Perspectivas no controle de formigas cortadeiras. Série Técnica IPEF, São Paulo. 1997;**11**(30):31-46

[13] Forti LC, Pretto DR, Nagamoto NS, Padovani CR, Camargo RS, Andrade APP. Dispersal of the delayed action insecticide sulfluramid in colonies of the leaf-cutting ant *Atta sexdens rubropilosa* (Hymenoptera: Formicidae). Sociobiology. 2007;**50**:1149-1163

[14] Britto JS, Forti LC, Oliveira MA, Zanetti R, Wilcken CF, Zanuncio JC, Loeck AE, Caldato N, Nagamoto NS, Lemes PG, Camargo RS. Use of alternatives to PFOS, its salts and PFOSF for the control of leaf-cutting ants *Atta* and *Acromyrmex*. International Journal of Research in Environmental Studies. 2016;**3**:11-92

[15] Mariconi FMA. As saúvas. São Paulo: Agronômica Ceres; 1970

[16] Camargo RS, Forti LC, Fujihara RT, Roces F. Digging effort in leaf-cutting ant queens (*Atta sexdens rubropilosa*) and its effects on survival and colony growth during the claustral phase. Insectes Sociaux. 2011;**58**:17-22

[17] Wilson EO. The Insect Societies. Cambridge: Belknap Press; 1971

[18] Corrêa MM, Silva PSD, Wirth R, Tabarelli M, Leal IR. How leaf-cutting ants impact forests: Drastic nest effects on light environment and plant assemblages. Oecologia, New York. 2010;**162**(1):103-115

[19] Bieber AGD, Oliveira MA, Wirth R, Tabarelli M, Leal IR. Do abandoned nests of leaf-cutting ants enhance plant recruitment in the Atlantic Forest? Austral Ecology, Malden. 2011;**36**:220-232

[20] Roces F, Hölldobler B. Leaf density and a trade-off between load-size selection and recruitment behavior in the ant *Atta cephalotes*. Oecologia. 1994;**97**:1-8

[21] Lopes JFS, Forti LC, Camargo RS. The influence of the scout upon the decision-making process of recruited workers in three *Acromyrmex* species. Behavioural Processes. 2004;**67**:471-476

[22] Roces F, Núnez JA. Information about food quality influences loadsize selection in recruited leaf-cutting ants. Animal Behavior, Canada. 1993;**45**(1):135-143

[23] Garrett R, Carlson KA, Goggans M, Nesson MH, Shepard CA, Schofield RMS. Leaf processing behaviour in Atta leafcutter ants: 90% of leaf cutting takes place inside the nest, and ants select pieces that require less cutting. Royal Society Open Science. 2016;**3**:150111. DOI: 10.1098/rsos.150111

[24] Sung GH, Hywel-Jones NL, Sung JM, Luangsa-Ard JJ, Shrestha B, et al. Phylogenetic classification of *Cordyceps* and the clavicipitaceous fungi. Studies in Mycology. 2007;**57**:5-59

[25] Butt T, Coates C, Dubovskiy I, Ratcliffe N. Entomopathogenic fungi: New insights into host–pathogen interactions. Advances in Genetics. 2016;**94**:307-364

[26] Ortiz-Urquiza A, Keyhani NO. Action on the surface: Entomopathogenic fungi versus the insect cuticle. Insects. 2013;**4**:357-374

[27] Alves SB. Controle microbiano de insetos. 4th ed. Piracicaba: FEALQ; 1998

[28] Travaglini RV, Forti LC, Arnosti A, Camargo RS, Silva LC, Camargo-Mathias MI. Mapping the adhesion of different Fungi to the external integument of *Atta sexdens rubropilosa* (Forel, 1908). International Journal of Agriculture Innovations and Research. 2016;**5**:118-125

[29] Lu H-L, St. Leger RJ. Insect Immunity to Entomopathogenic Fungi. College Park: University of Maryland; 2016

[30] Quiroz LJV. Factores de mortalidad natural de reinas de *Atta mexicana* (Fr. Smith) en el Estado de Morelos, México. In: Congresso Latinoamericano de Entomologia, 6, Mérida, Yucatán, 1996. Memorias. Mérida: Sociedad Mexicana de Entomologia; 1996. p. 56

[31] Alves SB, Sosa Gomez DR. Virulência do *Metharhizium anisopliae* (Metsch.) Sorok. e *Beauveria bassiana* (Bals.) Vuill. para castas de *Atta sexdens rubropilosa* (Forel, 1908). Poliagro. 1983;**5**(1):1-9

[32] Diehl-Fleig E, Da Silva ME, Bortolás EP. Emprego do fungo entomopatogênico *Beauveria bassiana* em iscas para o controle das formigas cortadeiras *Acromyrmex* spp. em floresta implantada de *Eucaliptus grandis*. Nova Prata: Congresso Florestal Estadual; 1992. pp. 1139-1150

[33] Kermarrec A, Febuay G, De Charme M. Protection of leaf-cutting ants from biohazards: Is there future for microbiological control? In: Lofgren CS, Vander Meer RK, editors. Fire Ants and Leaf-Cutting Ants – Biology e Management. Boulder: Westview Press; 1986. pp. 339-356

[34] Della Lucia TMC, editor. As formigas cortadeiras. Ed. Folha Nova de Viçosa; 1993. pp. 163-190

[35] Machado V et al. Reações observadas em colônias de algumas espécies de Acromyrmex (Hymenoptera-Formicidae) quando inoculadas com fungos entomopatogênicos. Ciência e Cultura, São Paulo. 1988;**40**(11):1106-1108

[36] Andrade APP. Comportamento forrageiro e aprendizado de operárias de Atta sexdens rubropilosa Forel, 1908 (Hymenoptera, Formicidae) em condições de campo e laboratório. MSc. Thesis. Botucatu, São Paulo, Brazil: Universidade Estadual Paulista; 1997. 100f

[37] Lucchese MEP. Desenvolvimento e avaliação de diferentes sistemas de inoculação do fungo entomopatogenico *Beauveria bassiana* em colônias de Atta *sexdens piriventris*. Resumo. p. 527. In: Congresso Brasileiro De Entomologia, 11, Encontro De Mirmecologia, 8; 1987

[38] Euzébio DE, Martins GF, Fernandes-Salomão TM. Morphological and morphometric studies of the antennal sensilla from two populations of *Atta robusta* (Borgmeier 1939) (Hymenoptera: Formicidae). Brazilian Journal of Biology. 2013;**73**(3):663-668. DOI: 10.1590/S1519-69842013000300026

[39] Slone JD, Pask GM, Ferguson ST, Millar JG, Berger SL, Reinberg D, Liebig J, Ray A, Zwiebel LJ. Functional characterization of odorant receptors in the ponerine ant, *Harpegnathos saltator*. PNAS. 2017;**114**(32):8586-8591. DOI: 10.1073/pnas.1704647114

[40] Zarzuela MFM, Leite LG, Marcondes JE, De carvalho campos AE. Entomopathogens isolated from invasive ants and tests of their pathogenicity. Psyche *(Stuttg)*. 2012;**2012**:1-9

[41] De Paula AR, Vieira LPP, DÁttilo WFC, Carneiro CNB, Erthal MJ, Brito ES, Silva CPP, Samuels RI. Patogenicidade e efeito comportamental de *Photorhabdus temperata* K122 nas formigas cortadeiras *Acromyrmex subterraneus subterraneus* e *Atta laevigata* (Hymenoptera: Formicidae). O Biológico. 2006;**68**:311-413

[42] Hazir S, Shapiro-Ilan DI, Bock CH, Hazir C, Leite LG, Hotchkiss MW. Relative potency of culture supernatants of Xenorhabdus and Photorhabdus spp. on growth of some fungal phytopathogens. European Journal of Plant Pathology. 2016;**146**:369-381

[43] Freitas ADG, Souza AQL, Maki CS, Pereira JO, Silva NM. Atividade antagonista de bactérias endofíticas de plantas da Amazônia contra o fungo simbionte *L. gongylophorus*, e dos fungos associados presentes nos ninhos de Atta sexdens. Sci Amazon. 2016;**5**:1-14

[44] Rocha SL, Evans HC, Jorge VL, Cardoso LAO, Pereira FST, Rocha FB, Barreto RW, Hart AG, Elliot SL. Recognition of endophytic *Trichoderma* species by leaf-cutting ants and their potential in a Trojan-horse management strategy. The Royal Society Publishing. 2017:1-14. DOI: 10.5061/dryad.0164h

[45] Currie CR, Stuart AE. Weeding and grooming of pathogens in agriculture by ants. Proceedings of the Royal Society of London. Series B. 2001;**268**:1033-1039

[46] Currie CR, Wong B, Stuart AE, Schultz TR, Rehner SA, Muller UG, Sung GH, Spatafora JW, Straus NA. Ancient tripartite coevolution in the attine ant-microbe symbiosis. Science. 2003;**299**:386-388. DOI: 10.1126/science.1078155

[47] Dornelas ASP, Sarmento GR, Nascimento MO, Oliveira CA, Souza DJ. *Atta sexdens* worker ants treated with the immunosuppressant Sandimmun Neoral to *Metarhizium anisopliae*. Pesquisa Agropecuária. 2017. DOI: 10.1590/s0100204x2017000200008

[48] Lofgren CS, Vander Meer RK. Fire Ants and Leaf-Cutting Ants – Biology e Management. Boulder, CO: Westview Press; 1986. pp. 339-356

[49] Galvanho JP, Carrera MP, Moreira DDO, Erthal M, Silva CP, Samuels RI. Imidacloprid inhibits behavioral defences of the leaf-cutting ant *Acromyrmex subterraneus* subterraneus (Hymenoptera: Formicidae). The Journal of Insect Behavior. 2013;**26**:1-13

[50] Gilljam JL, Leonel J, Cousins IT, Benskin JP. Is ongoing Sulfluramid use in South America a significant source of Perfluorooctanesulfonate (PFOS)? Production inventories, environmental fate, and local occurrence. Environmental Science & Technology. 2016;**50**(2): 653-659

[51] Castilho AMC, Fraga ME, Aguiar-Menezes EL, Rosa CAR. Seleção de isolados de *Metarhizium anisopliae* e *Beauveria bassiana* patogênicos a soldados de *Atta bisphaerica* e *Atta sexdens rubropilosa* em condições de laboratório. Ciência Rural. 2010;**40**:1243-1249

[52] Catalani GC, Sousa KKA, Stefanelli LEP, Travaglini1 RV, Forti LC. Estudos Com Plantas Inseticidas Para O Controle De Formigas Cortadeiras. In: Baldin ELL, Kronka AZ, Silva IF, editors. Inovações em manejos fitossanitários. 1st ed. Botucatu: Fundação de Estudos e Pesquisas Agrícolas e Florestais; 2017

[53] Loreto RG, Hughes DP. Disease dynamics in ants: A critical review of the ecological relevance of using generalist fungi to study infections in insect societies. Advances in Genetics. 2016;**94**:287-306. DOI: 10.1016/bs.adgen.2015.12.005

[54] Rodrigues A, Bacci MJR, Mueller UG, Ortiz A, Pagnocca FC. Microfungal "weeds" in the leaf-cutter ant symbiosis. Microbial Ecology. 2008;**56**:604-614. DOI: 10.1007/s00248-008-9380-0

[55] Pagnocca FC, Carreiro SC, Bueno OC, Hebling MJA, Silva OA. Microbiological changes in leaf-cutting ant nests fed on sesame leaves. Journal of Applied Entomology. 1996;**120**:317-320

[56] Cardoso SRS. Morfogênese de ninhos iniciais, mortalidade de Atta spp. (Hymenoptera, Formicidae) em condições naturais e avaliação da ação de fungos entomopatogênicos. 100f. Tese Doutorado em Agronomia (Proteção de Plantas) – UNESP; 2010

[57] Poulsen M, Bot ANM, Nielsen MG, Boomsma JJ. Experimental evidence for the costs and hygienic significance of the antibiotic metapleural gland secretion in leaf-cutting ants. Behavioral Ecology and Sociobiology. 2002;**52**:151-157. DOI: 10.1007/s00265-002-0489-8

[58] Do Nascimento RR, Schoeters E, Morgan ED, Billen J, Stradling DJ. Chemistry of metapleural gland secretions of three attine ants, Atta sexdens rubropilosa, Atta cephalotes, and *Acromyrmex octospinosus* (Hymenoptera: Formicidae). Journal of Chemical Ecology. 1996;**22**(5):987-1000

[59] Fernandez-Marin H, Zimmerman JK, Rehner SA, Wcislo WT. Active use of the metapleural glands by ants in controlling fungal infection. Proceedings of the Royal Society B. 2006;**273**:1689-1695. DOI: 10.1098/rspb.2006.3492

[60] Lacerda FG, Della Lucia TMC, Desouza O, Pereira OL, Kasuya MCM, De Souza LM, Couceiro JC, De Souza DJ. Social interactions between fungus garden and external workers of Atta sexdens (Linnaeus) (Hymenoptera: Formicidae). Italian Journal of Zoology. 2014;**81**:298-303. DOI: 10.1080/11250003.2014.911369

[61] Oi DH, Briano JA, Vallesa SM, Williams DF. Transmission of *Vairimorpha invictae* (Microsporidia: Burenellidae) infections between red imported fire ant (Hymenoptera: Formicidae) colonies. Journal of Invertebrate Pathology. 2005;**88**:108-115. DOI: 10.1016/j.jip.2004.11.006

[62] Travaglini RV, Stefanelli LEP, Arnosti A, Camargo RS, Forti LC. Isca Encapsulada Atrativa Visando Controle Microbiano De Formigas Cortadeiras. Tekhne E Logos. 2017;**8**(2): 100-110

[63] Travaglini RV. Bases Para O Controle Microbiano De Formigas Cortadeiras. 93f. Tese (Doutorado em Proteção de Plantas) Universidade Estadual Paulista, Faculdade de Ciências Agronômicas, Botucatu; 2017

[64] Drake VA, Reynolds DR. Radar Entomology: Observing Insect Flight and Migration. Wallingford: CABI; 2012

[65] Favoreto AL, Puretz BO, De Souza AR, Neto HC, Bezerra NS Jr, Wilcken CF. Principais Técnicas De Monitoramento De Insetospraga. In: Baldin ELL, Kronka AZ, Silva IF, editors. Inovações Em Manejos Fitossanitários. 1st ed. Botucatu: Fundação de Estudos e Pesquisas Agrícolas e Florestais; 2017

[66] Arnosti A, Delalibera I Jr, Conceschi MR, d'Alessandro CP, Travaglini RV. Camargo-Mathias MI. Interactions of adjuvants on adhesion and germination of *Isaria fumosorosea* on *Diaphorina citri* adults Scientia Agricola; 2018

[67] Cremer S, Armitage SAO, Schmid-Hempel P. Social immunity. Current Biology. 2007; **17**:R693eR702

[68] Billen J. A importância de glândulas exócrinas na sociedade de insetos. In: Vilela EF et al., editors. Insetos Sociais: da Biologia à Aplicação. Viçosa: UFV; 2008. pp. 87-92

[69] Tranter C, Hugues WO. Acid, silk and grooming: Alternative strategies in social immunity in ants? Behavioral Ecology and Sociobiology. 2015;**69**(10):1687

[70] Lopez E, Orduz S. *Metarhizium anisopliae* and *Trichoderma viride* for control of nest of the fungus-growing ant, *Atta cephalotes*. Biological Control. 2003;**27**:194-200

[71] Man TJB, Stajich JE, Kubicek CP, Teiling C, Chenthamara K, Atanasov AL, Druzhinina IS, Levenkova N, Birnbaum SSL, Barribeau SM, Bozick BA, Suen G, Currie CR, Gerardo NM. Small genome of the fungus Escovopsis weberi, a specialized disease agent of ant

agriculture. Proceedings of the National Academy of Sciences of the United States of America. 2016;**113**:3567-3572. DOI: 10.1073/pnas.1518501113

[72] Masiulionis VE, Cabello MN, Seifert KA, Rodrigues A, Pagnocca FC. *Escovopsis trichodermoides* sp. nov., isolated from a nest of the lower attine ant *Mycocepurus goeldii*. Antonie Van Leeuwenhoek. 2015;**107**:731-740. DOI: 10.1007/s10482-014-0367-1

[73] Rodrigues A, Pagnocca FC, Bacci MJ, Hebling MJA, Bueno OC, Pfenning LH. Variability of non-mutualist fungi associated with Atta sexdens rubropilosa nests. Folia Microbiologica. 2005;**50**:421-425